Materials Horizons: From Nature to Nanomaterials

Series Editor

Vijay Kumar Thakur, School of Aerospace, Transport and Manufacturing, Cranfield University, Cranfield, UK

Materials are an indispensable part of human civilization since the inception of life on earth. With the passage of time, innumerable new materials have been explored as well as developed and the search for new innovative materials continues briskly. Keeping in mind the immense perspectives of various classes of materials, this series aims at providing a comprehensive collection of works across the breadth of materials research at cutting-edge interface of materials science with physics, chemistry, biology and engineering.

This series covers a galaxy of materials ranging from natural materials to nanomaterials. Some of the topics include but not limited to: biological materials, biomimetic materials, ceramics, composites, coatings, functional materials, glasses, inorganic materials, inorganic-organic hybrids, metals, membranes, magnetic materials, manufacturing of materials, nanomaterials, organic materials and pigments to name a few. The series provides most timely and comprehensive information on advanced synthesis, processing, characterization, manufacturing and applications in a broad range of interdisciplinary fields in science, engineering and technology.

This series accepts both authored and edited works, including textbooks, monographs, reference works, and professional books. The books in this series will provide a deep insight into the state-of-art of Materials Horizons and serve students, academic, government and industrial scientists involved in all aspects of materials research.

More information about this series at http://www.springer.com/series/16122

A. K. Soni · P. Nema

Limestone Mining in India

Springer

A. K. Soni
CSIR-Central Institute of Mining and Fuel Research (CSIR-CIMFR)
Nagpur, Maharashtra, India

P. Nema
CSIR-National Environmental Engineering Research Institute (CSIR-NEERI)
Nagpur, Maharashtra, India

ISSN 2524-5384 ISSN 2524-5392 (electronic)
Materials Horizons: From Nature to Nanomaterials
ISBN 978-981-16-3559-5 ISBN 978-981-16-3560-1 (eBook)
https://doi.org/10.1007/978-981-16-3560-1

This Springer imprint is published by the registered company Springer Nature Singapore Pte Ltd.
The registered company address is: 152 Beach Road, #21-01/04 Gateway East, Singapore 189721, Singapore

Foreword

Mining is an old activity taken up by humans at the inception of civilization. For ages, minerals have formed the essential component of all manmade valuable and usable products. Upon examination of the mining activities of the present day, particularly limestone mining on the Indian Peninsula, it is clear that several transformations in the limestone mining sector from those early days to the present have taken place, mainly for the better. But the industry continues to grow and change. My long association with geology and limestone mining, in particular for the Indian cement sector, has taught me that mineral resource-related commercial activities are now in a transitional phase for both the industry and its entrepreneurs. The legislative framework governing the industry is also being upgraded and strengthened. Modernization within the industry will help to make the 'unknown' known. This newly available knowledge may help to yield a sizable amount of money from mineral sector development, which will contribute immensely to the GDP growth. Undoubtedly, the limestone mining sector will have a significant contribution to the GDP growth as well.

Limestone, as a mineral, has undergone many changes and transformations to become a sedimentary rock and the principal raw material for cement manufacturing. The focus of this book is on the ways of mining such a valuable resource. Open surface mines, the low cost of the mineral, the large variations in grades and constitution, the locations in different geographical areas, field conditions and topography, and the overall mining conditions are just some of the limestone excavation topics addressed by the authors in this text. In the twenty-first century, limestone mines are now utilizing new, modern types of machinery and equipment for faster excavation. By and large, the 'conventional methods' of mining remain in vogue. But this current information era and recent advancement in communication techniques have transformed many limestone mining-related activities.

Limestone Mining in India is a book written by two highly experienced authors and will certainly make a valuable and impressive contribution to the monograph series on the limestone mineral, focusing exclusively on its mining aspects. This text will help to bridge the current knowledge gap that exists in the limestone sector to an extent that its readers will be able to obtain the details of nearly all major aspects of limestone mining activities. I am confident that this reference book will explain how mining and the protection of the environment can be integrated and improved further to make mining practices green and eco-friendly. The state-of-the-art knowledge described in the case studies that appear in this text delineates the existing practices and explains the various practical aspects as applied in the limestone mining industry, a part of the commercial Indian mineral industry.

In this Foreword, I urge the mine operators of limestone mines, particularly those from the cement sector, to take the benefit of this book in their day-to-day workings. At all levels, simple applied concepts and contents discussed in this text will be helpful for the ROM production and environmental management of the operational mines.

I have noted that this publication is a contribution of CSIR scientists, one from mining research and the other from environmental research. The authors' efforts here have brought laurels and yet another feather in the cap of CSIR research efforts. I hope that the *science of mining* and scientific knowledge of minerals, the environment, and engineering will grow through the sincere efforts of these authors in the preparation of this book. This book is written with a focus on mining on the Indian continent, but be aware that this particular topic will draw the attention of other developing countries, namely Nepal, Bhutan, Bangladesh, Afghanistan, and the like. These developing nations possess good quality limestone resources in ample quantity and scientist and engineers from these nations could benefit greatly from reading this text.

Plenty of limestone resources are still hidden and remain unexploited; hence, the developmental journey of limestone mining will continue well into the future. My kudos for the authors' effort to develop and contribute to this sector both scientifically and technically.

I congratulate both the authors for their praiseworthy work. Keep it up!!

March 2021

Yellarthi Raghvendra Rao
Former Joint President, Ambuja Cement
Hyderabad, India

Preface

Limestone Mining in India is a comprehensive presentation of the scientific and technological aspects of limestone mining. It emphasizes how limestone is excavated in India from surface mines under different geomining conditions. Existing practices are described through case studies delineating and explaining the essentials of the mine, mine production, environmental management, and different commercial aspects as applied for the limestone mining industry, which is a part of the Indian mineral industry.

In India, the mineral limestone is covered in both major mineral and minor mineral categories, depending on its use. Deposits are found in different geological areas across the length and breadth of the country in several states with ample reserves for commercial exploitation.

This reference book contains topical issues related to the quarrying of limestone using open cast mining. The critiques and gaps in limestone mining and environmental practices provide practical solutions for the trouble shooting of key technical problems (e.g. improving productivity through blast optimization) that are discussed at length. New technological aspects covering how these solutions can be scientifically planned to make the project more efficient, practical, and cost-effective is the central theme of this book. As a result, the reader can come to know what needs to be done in future for limestone mining practices. This book maintains a balance between scientific information, technology, and general practices in the present and for the future, although many of the described areas are very vast and elaborate. More than three decades of the rich experience of the authors in mining, environmental subjects, and analysis based on integrated solutions, new ideas, and eco-friendly approaches enrich the overall technical content of this book. Engineers, scientists, researchers, and students should be (in the authors' opinion) the direct beneficiaries of the work presented in this book.

Limestone excavation and its related aspects as major mineral are well documented in the form of technical papers, technical reports, etc., but the necessity for exclusive documentation of limestone mining in book form left a gap the authors have attempted to fill. Our focus has been kept on all major mining and environmental issues of limestone mines. The book describes in what manner the mining processes can be planned or designed to ensure that the produced limestone meets the market demand.

This book also explains and emphasizes the environment-oriented development of limestone.

Limestone may be a mineral of low cost, but it is an important mineral in the industry right from the time Portland Cement was patented by Joseph Aspdin in 1824. The importance of limestone has remained constant always and never came down, which prompted us to write this book for the Indian mineral sector. Limestone is being mined by various industrial sectors, namely cement, chemical, and steel. However, the cement sector consumes nearly 80 to 85% of its production, which is the highest among all sectors, as raw material feed for manufacturing of different varieties of cement. The authors believe that this book will contribute its role as a purveyor of scholarly scientific contribution. The information contained herein is indeed for the limestone mining sector but is also equally applicable to other important issues of our concern—environmental protection and preservation of the Earth's ecosystem in which we all live. Local and cost-effective solutions for mining and related environmental problems and issues also have been discussed. Applications of modern technologies and their entailing benefits are useful for both industry and society in particular through mineral exploitation. The limestone mining sector in India is heading towards holistic growth and should follow or adopt the green mining and sustainable development routes. Being a forward-looking sector, the mining industry has a tremendous scope for future improvements and advancements.

Nagpur, India

A. K. Soni
P. Nema

Acknowledgements

We would like to place on record and duly acknowledge all of those companies whose mine's names are listed in this book: M/s Lafarge Umiam Mining Private Limited (LUMPL), Shillong, Meghalaya; M/s Trinetra Cements Limited (India Cements) Banswara, Rajasthan; and M/s India Cement, Chennai; M/s Manikgarh Cement, Gadchandur, Maharashtra; M/s UltraTech Cement, Awarpur Cement Works (ACW), Chandrapur, Maharashtra; M/s UltraTech Cement, Gujarat Cement Works (GCW), Gujarat and Narmada Cement Jafarabad Works (NCJW), Gujarat. Without the help of these companies, it would not have been possible for us to write case studies and describe various ideas about limestone mining and its related aspects.

We are extremely thankful to Sri D. Mahule of UltraTech Cement; Sri Bharat Gokharu (NCJW); Dr. R. K. Mishra (ACW /NCJW); Shri Adesh Mishra (Lafarge Cement); Shri Sanjay Kumar (LUMPL); (Late) Shri S. K. Jain; Shri V. C. Joshi (India Cement); and Shri Jayant Mohgaonkar (Orient Cement) for their support and encouragement. The authors thankfully acknowledge Ms. Mrualini V. Khond, Geologist of the Geological Survey of India (GSI), Hyderabad, for permitting us to make use of data, which was a part of her Ph.D. research work, for in-depth analysis. The cooperation of Shri R. K. Udge, Ex-Head (Mines), Manikgarh Cement, Shri Rajesh Sambhare, former Mines Manager, Naokari Limestone Mine, and the assistance provided by the Deputy Director and Senior chemist, Ground Water Survey and Development Agency (GSDA), Nagpur, who helped in the field study and water quality analysis of the Case Study 5, is duly acknowledged.

The authors are extremely thankful to the Director, CSIR-CIMFR, Dhanbad, and Director, CSIR-NEERI, Nagpur for their support and encouragement during writing of this book. Indeed, direct and indirect organizational help taken from time to time in some component of the research work reported in this book is worthy of acknowledgement. During our stay in our respective institutes over the years, we have learned a great deal and we hope that we have disseminated our collective knowledge in this book.

In this sojourn, several practical difficulties were encountered, however they could be sorted out and overcome with the help of our colleagues both in the office and at the mine sites in the field. Since it is not possible to name them individually for want of space, we thankfully acknowledge all of them.

Last but not the least, we thank all those who rendered their help and support directly or indirectly in shaping this work in a book form.

A. K. Soni
P. Nema

Contents

About the Authors

Dr. A. K. Soni graduated in Mining Engineering from Ravishankar University, Raipur, India. He completed his postgraduate degree from Birla Institute of Technology and Science (BITS), Pilani and Ph.D. in Environmental Science and Engineering from Centre of Mining Environment, Indian School of Mines (ISM), Dhanbad (now IIT-ISM) in 1998. He is working as the Chief Scientist at the CSIR-CIMFR Nagpur Research Centre and is engaged in research in mine environment and allied areas. His research area of interest is geo-hydrological problems related to mines. He has been associated with the development of eco-friendly techniques of mineral extraction and also engaged in research on environmental management in fragile/sensitive areas with particular reference to mining operations and hill type areas (Himalaya). He developed an Environmental Degradation Index (EDI) for application in mining in ecologically fragile areas and has special interest in policy issues on mining and environment. He is associated with large number of mining and tunnelling projects in India and invited by Indian universities to deliver lectures. He has more than 115 technical publications on environment related topics in national and international journals, conference proceedings, and workshops. He has handled over 100 R&D projects in the capacity of principal coordinator and investigator. To his credit, Dr. Soni has authored a book on 'Mining in the Himalayas- An Integrated Strategy', published by CRC Press, Taylor & Francis (2017) and is an Academic Editor for an open access book titled "Mining Techniques: Past, Present and Future" published by Intech Open, London (2021).

Dr. P. Nema obtained his BE and ME in Civil Engineering with specialization in Public Health Engineering degrees from Jabalpur University, Madhya Pradesh in 1971 and 1975 respectively. He was awarded the Netherlands Government Fellowship in 1979 and completed his post graduate Diploma in Sanitary Engineering from the International Institute for Hydraulic and Environmental Engineering (IHE), Delft, The Netherlands in 1980. He obtained Ph.D. in Environmental Engineering from Indian Institute of Technology (IIT) Roorkee in 1998. He had a long association with National Environmental Engineering Research Institute (NEERI), Nagpur as a scientist from 1975 to 2008 and continued post-superannuation as Project Advisor until 2013. He has over three decades experience in Research & Development in Water,

Wastewater Treatment/Management, Air Quality Management and Environmental Impact Assessment of Developmental Projects in different sectors. He has prepared more than sixty project reports and published more than 50 research papers in journals of both national and international repute, and conference proceedings. His academic interests included working as visiting faculty to PG programs in Environmental Engineering in Indian Universities. He has also been a reviewer for several prominent journals in the fields of bio-resource technology, environmental engineering and desalination and water treatment.

Abbreviations

ACC	Associated Cement Company Limited
AMD/ARD	Acid mine drainage or / Acid rock drainage
AQ	Air quality
BGL/AGL	Below ground level/above ground level
BIS	Bureau of Indian Standard (Formerly ISI)
CGWA	Central Ground Water Authority
CGWB	Central Ground Water Board
CIMFR	Central Institute of Mining and Fuel Research (CIMFR) *(formerly, Central Institute of Mining Research or CMRI)*
CMRI	Central Mining Research Institute
CPCB	Central Pollution Control Board
CSM	Continuous surface miners
EFA	Ecologically fragile areas
EIA	Environment Impact Assessment
EIS	Environmental Impact Statement
EMP	Environment Management Plan
EPA	Environment Protection Act, 1986
EPR	Environment Protection Rules, 1986
GDP	Gross domestic product
GOI	Government of India
GSI	Geological Survey of India
HEMM	Heavy earth-moving machinery
IBM	Indian Bureau of Mines
MECL	Mineral Exploration Corporation Limited
MOEFCC	Ministry of Environment, Forest Wildlife and Climate Change (Government of India)
MT	Million Tonnes
NGO	Nongovernmental organization
RL	Reduced level (elevation with reference to a datum)
ROM	Run of mine
SSM	Small-scale mines
Tpa	Tonnes per annum

TPD/tpd	Tonnes per day
TPY/tpy	Tonnes per year
TM	Trademark
US-EPA	United States Environmental Protection Agency
WHO	World Health Organization
WQ	Water quality

List of Figures

List of Tables

List of Boxes

Chapter 1
About Limestone

Mining was a traditional skill in older days and considered to be an art; however, technological advances have slowly turned this art into an applied skill science. The long journey from *art* to *expertise science* has brought many technical upgrades to the mineral and mining industries that have benefitted society. India is endowed with large, exploitable limestone reserves. The available data shows there are more than 2100 operating limestone mines of big and small sizes are operating in India across many states and union territories [1]. The positive growth profile of limestone mining and connected ancillary industrial activities is shaping the current trend of limestone mining in the twenty-first century. In India, limestone mining is carried out only through the opencast method. This chapter will briefly provide information about limestone production, uses, geological occurrences, characteristics, etc. This will prepare the groundwork to create a base for various mining and environment issues. The descriptions will prove advantageous for understanding all related aspects of limestone mining in India. In subsequent chapters, through review and discussion of different case studies, the reader will understand the different terrains and conditions, alternate methods of excavation, equipment, and machinery used in Indian limestone mines. The industries consuming limestone will form the medium of description for this book.

1.1 Limestone Production

Exploration and exploitation of limestone deposits have geared momentum in India's post-Independence period because of intensified industrial demand. The decades from 1990 to 2000 were important for the commercial exploitation of limestone when the Indian economy was liberalized. Cement, iron and steel, sugar refractories, paints, and chemical industry witnessed all round growth, boosting the requirement for limestone. Different grades of limestone (cement grade, steel grade, chemical grade, etc.) were in demand and explored in different states all over the country. With

A. K. Soni and P. Nema, *Limestone Mining in India*, Materials Horizons: From Nature to Nanomaterials, https://doi.org/10.1007/978-981-16-3560-1_1

Table 1.1 Production of limestone in India (by state)

S. No	Name of state	2014–15	2015–16	2016–17
1	Andhra Pradesh	34,676	32,579	35,251
2	Assam	665	777	1594
3	Bihar	473	459	190
4	Chhattisgarh	23,588	27,667	31,919
5	Gujarat	26,010	25,622	24,923
6	Himachal Pradesh	12,710	12,390	11,009
7	Jammu & Kashmir	130	1232	825
8	Jharkhand	792	1076	1146
9	Karnataka	24,008	27,062	29,784
10	Kerala	511	487	376
11	Madhya Pradesh	39,530	39,430	35,843
12	Maharashtra	12,085	13,036	12,119
13	Meghalaya	3691	3834	5103
14	Odisha	3409	4532	4751
15	Rajasthan	61,844	67,336	67,078
16	Tamil Nadu	22,227	23,008	23,840
17	Telangana	23,972	23,878	24,789
18	Uttar Pradesh	2952	2596	2656

Note All values are in '000 tonnes
Source Indian Bureau of Mines, Nagpur [9]

this accelerated industrial demand of limestone as a raw material, the production of limestone is also increased. The production in India in various states for recent years is given in Table 1.1.

Cement plants of both large and small capacity that were principal consumers of limestone are since operative in several states of India (Table 1.2), a persistent demand of limestone always existed. Consequently, limestone deposits were explored, and a number of new mines were added to the mining inventory.

Cement grade high-quality limestone is easily and abundantly available across India. Limestone, being an essential constituent of all types of cement manufacturing, is largely consumed in the cement industry to the extent of nearly 85% or more of its production, which is the highest from among all other limestone-consuming industries [10]. In addition to limestone, some other minerals (coal, clay, gypsum) that are used as additives in the cement industry are also excavated from mines. Other additives include fly ash, which is generated as a waste product from the generation of electricity, and slag residues, which come from the blast furnaces of steel plants. However, none of these additive inputs currently place any constraint in terms of availability or quality. Therefore, the whole limestone sector (production, consumption, demand and supply, and raw material uses) has a bright future in India.

Table 1.2 Cement plants of India with installed capacity above and below 0.5 MT

Name of state	Installed capacity above 0.5 MT		Installed capacity less than 0.5 MT	
	No. of plants	Percentage share	No. of plants	Percentage share
Andhra Pradesh	32	21.33	3	9.09
Assam	1	0.67	1	3.03
Chhattisgarh	8	0.67	2	6.06
Gujarat	11	9.09	3	9.09
Haryana	2	3.03	1	3.03
Himachal Pradesh	2	3.03	1	3.03
Jharkhand	3	2.00	2	6.06
Karnataka	8	5.33	3	9.09
Madhya Pradesh	11	7.33	1	3.03
Maharashtra	9	6.00	1	3.03
Meghalaya	1	0.67	3	9.09
Punjab	2	1.33	1	3.03
Rajasthan	18	12.00	2	6.06
Orissa	4	2.67	–	–
Delhi	–	–	1	3.03
Jammu & Kashmir	–	–	1	3.03

Note Installed capacity of various plants has been upgraded periodically
Source Mukhopadhyay [2]

Most of the functional captive limestone mines are owned by the private sector. *Captive mines* are those which are owned by a company for meeting its own requirement of different minerals and will not sell its products in the open market, e.g. mostly cement manufacturing companies have their own limestone mines to meet their lime requirement (as raw material). Most limestone mines are assigned the status of 'captive mines', and very few are owned by the government or public sector [e.g. Cement Corporation of India (CCI) owns plant and captive mines at Bokajan (Assam) and Rajban (H.P.)]. All of these mines had adopted modern full- to semi-mechanized methods of mining and a dry process of cement manufacturing using limestone as the principal raw material.

1.2 Limestone Uses and Consuming Industries

Important mineral economies in the world, such as South Africa (largest), Australia, the United States of America (USA), the United Kingdom, and Canada, have thriving mineral-based industries that are dominated by a handful of major private mining companies. Because of significant mineral demands, its rich mineral resources, and the related stimulus of its overall economic growth, India has emerged as a new industrialized economy of Asia. Indian mining and mineral sectors contribute significantly to its GDP by producing 95 principal minerals (four fuel, five atomic, ten metallic, 21 non-metallic, and 55 other minor minerals) from across the 28 states and eight union territories. Limestone is in the non-metallic category (see Table 1.3). In India, limestone is excavated through 2117 leases of different sized mines covering a net geographical area of about 160,000 ha [1]. There is a large availability of both major and minor categories of limestone used for reserves, exploration, and the possibility of exploitation. The industrial need and consumption of limestone are also very large. Limestone's reserve evaluation has been estimated by various exploration agencies for different categories. As per the United Nations (UN) norm, these are summarized in Table 1.4. The reserves and its regional distribution of limestone in India give a clear indication that it is the most abundantly found mineral of the nation (see Table 1.5). Because limestone is found nearer to surface, easy mining conditions and abundant reserves make it amenable for successful commercial exploitation by surface mining.

The total reserves/resources of limestone of all categories and grades as per *national mineral inventory* (NMI) data based on the United Nations Framework Classification for Resources (UNFC) system as on 2015 have been estimated at 203,224 million tonnes, of which 16,336 MT (about 8%) are placed under the reserves category. Karnataka is the leading state having 27% of the total resources. This is followed by Andhra Pradesh and Rajasthan (12% each), Gujarat (10%), Meghalaya (9%), Telangana (8%), Chhattisgarh, and Madhya Pradesh (5% each), and the remaining 12% by other states. Cement grade has a leading share of about 70%, followed by unclassified grades at 12%, and BF grade at 7%. The remaining 11% are various different grades. The total resources of marl (a limestone variety found mainly in the Gujarat coast) of all categories and grades have been estimated at 135.56 MT of which 123.86 million tonnes (91%) are under the reserves category (as on 2015) as per NMI data based on the UNFC system.

Limestone uses are plenty, e.g. for cement manufacturing as raw material feed to kiln (see Table 1.6); in metallurgical industry as flux; in foundries as feed to an LD furnace; in the iron and steel industries as feed to a blast furnace and steel melting shops [3]; for manufacturing of bleaching powder; in the sugar industry for the cleaning of cane juices; in the glass-making industry as a colouring agent; in the fertilizer industry for the manufacture of calcium ammonium nitrate; as refractory raw material in LD furnaces and for fettling purposes in steel melting shops (SMS); for making calcium carbide; for the manufacture of caustic soda and soda ash; as fillers or extenders in the chemical industry; and to treat acidic water as an ameliorant. Besides

Table 1.3 Minerals in India

(a) List of major minerals in India (by category)[a]

Fuel minerals (4)	**Atomic minerals (5)**	
– Coal – Lignite – Petroleum (crude) – Natural Gas	– Uranium – Monazite – Ilmenite – Rutile – Zircon	
Metallic minerals (10)	**Non-metallic minerals (21)**	
– Bauxite – Chrome ore – Copper ore – Gold – Iron ore – Lead – Manganese ore – Zinc – Tin – Silver (by-product)	– Asbestos – Apatite – Phosphorite rock phosphate – Diamond – Garnet – Graphite – Kyanite – Vermiculite – Wollastonite – Siliceous earth	– Limestone – Limeshell – Magnesite – Sillimanite – Selenite – Fluorite – Flint stone – Marl – Moulding sand – Sulphur (by-product) – Salt

(b) List of minor minerals excavated in India[a, b]

Notification	**Minor minerals (55)**
Section 3(e) of MMDR Act, 1957	• Building stone • Gravel • Ordinary clay • Ordinary sand other than sand used for prescribed purpose (Rule 70 of Mineral Concession Rules, 1960) – Refractory and manufacture of ceramic – Metallurgical purposes – Optical purposes – Purposes of stowing in coal mines – Manufacture of silvicrete cement – Manufacture of sodium silicate – Manufacture of pottery and glass
Notification No. M.II 159(18)/54-A.II dated Notification No. M-II-169(40)/58 dated 01/06/1958	– Boulder – Shingle – Chalcedony pebbles used for ball mill purposes only – Limeshell, kankar, and limestone (used in kilns for manufacture of lime used as building material) – Murrum – Brick earth – Fuller's earth – Bentonite – Road metal – Reth-matti – Slate (when used for building material) – Shale (when used for building material)

(continued)

Table 1.3 (continued)

List of minor minerals covered under the jurisdiction of the state governments		
Notification No. M.II/159(6)57 dated 03/09/59	– Marble	
Notification No. M. II 1(1)/63-dated 25/02/65	– Stones used for making household utensils – Quartzite – Sandstone (when used for purposes of building or for making road metal and household utensils)	
Notification No.1 (31)/65-M.II dated 28/01/67	– Saltpeter	
GSR 95(E) dated 03/02/2000	– Ordinary earth (used for filling or levelling purposes in connection with embankments, roads, railways, and building)	
GSR 333 dated 10/02/2015	– Agate – Ball clay – Barytes – Calcarious sand – Calcite – Chalk – China clay – Clay (others) – Corundum – Diaspore – Dolomite – Dunite or Pyroxenite – Felsite – Felspar – Fireclay	– Fuschite – Quartzite – Gypsum – Jasper – Kaolin – Laterite – Lime Kankar – Mica – Ochre – Pyrophyllite – Quartz – Quartzite – Sand (others) – Shale – Silica sand – Slate, and – Steatite, talc, or soapstone

[a] *Source* IBM, Nagpur, India

[b] *Note* Unless stated all dates mentioned in the table are as per dd/mm/yy pattern, as followed In India

these uses, limestone (together with dolomite) being used as a building material and in agricultural industries is quite common and popular too.

Important limestone-consuming industries in descending order are cement; chemical; steel and foundry; sugar; paper; glass; paint and dyes; and fertilizer. Limestone in various forms is also used quite commonly in building and road construction.

Table 1.4 Reserves of limestone in India (in '000 tonnes as in 2015)

Grade/state	Reserves			
	Proved	Probable		Total
	STD 111	STD121	STD122	
All India (total by grades)	9,438,939	3,015,917	3,880,897	16,335,753
Chemical	184,411	98,399	95,562	378,372
SMS (OH)	135,571	853,518	10,146	999,235
SMS (LD)	2636	182	584	3402
SMS (OH and LD mixed)	–	–	–	–
BF	247,462	44,404	51,201	343,068
SMS and BF mixed	40,226	101,941	27,728	169,894
Cement (Portland)	8,373,610	1,693,372	3,549,049	13,616,030
Cement (white)	133	23	115	270
Cement (Portland and white)	1776	–	930	2706
Cement (blendable/beneficiable)	183,933	51,087	64,749	299,769
BF and cement mixed	49,731	208	35,456	85,394
Limestone (chemical and paper grade)	2207	–	273	2479
Paper	25,551	–	–	25,551
Others	43,906	41,787	7861	93,555
Unclassified	138,164	108,746	36,731	283,642
Unknown	9623	22,250	513	32,385

Source http://ibm.nic.in/writereaddata/files/01092019132520Limestone%202017.pdf; [Indian Minerals Year Book, 2017, Vol. III: Reviews on Minerals, IBM, Nagpur, pp. 18–3
Notations: *SMS* steel melting shop; *BF* blast furnace; *OH & LD* open hearth and Linz & Donawitz—a process of steel making

1.2.1 Limestone as Major and Minor Minerals

Limestone in India is treated both as major mineral and minor mineral according to its use in industry. When used for cement, steel, glass, chemical, and other major industries in bulk quantity, it is a *major mineral* for which all types of industrial cess, taxes, etc., are applied. When the same limestone is used in kilns for calcium oxide or calcium carbide production, it is treated as a *minor mineral* and exempted from many taxes. As increased demand for limestone mineral has made higher production capacities necessary, such classification of major and minor was thought of. Accordingly, 'lime kankar' (a mineral with significant calcium content) has been brought in calcareous mineral category and declared as *minor mineral* vide Governmen of India notification S.O. 423 (E) dated 10/02/2015. It needs mention here that all producers of minor minerals file their *annual returns* of the production directly to the respective states and not to Indian Bureau of Mines (IBM), in contrary to the major minerals for which IBM is the nodal agency.

Table 1.5 Limestone reserves

(a) Region-wise distribution in India[a]

Name of region	**Gross** (in million tonnes and percentage)	**Proved** (in million tonnes)	**Remarks**
	97,430.45 MT	63,759.1 MT	Data shown here is as on March 2006
• Southern region	• 45.00%		
• Northern region	• 21.84%		
• Western region	• 12.34%		
• Eastern region	• 15.82%		
• Central region	• 03.64%		

(b) Distribution in India as per UNFC system[b]

Nomenclature	**UNFC designations**	**Quantity ('000 metric tons)**
Proved	111	8,978,583
Probable	121	3,650,576
Probable	122	2,297,234
Feasibility status	211	1,827,583
Prefeasibility status	221	3,739,470
Prefeasibility status	232	6,309,489
Measured	331	6,858,999
Indicated	332	22,040,640
Inferred	333	124,835,558
Reconnaissance status	334	4,396,981
Total	–	184,935,113

[a]*Source* CSIR-CIMFR [4]
[b]*Source* Chatterjee [5]

Table 1.6 Cement grade limestone

Chemical composition of constituent element of limestone	Permissible/acceptable range for cement industry (in %, max/min)
CaO	42 (minimum); 42–45 (permissible)
MgO	4 (maximum)
SiO_2	12–16 (maximum)
Al_2O_3	2–4
Fe_2O_3*	1–2
Mn_2O_3	3 (maximum)
TiO_2	1 (maximum)
P_2O_5	1 (maximum)
Chlorides	0.015
PbO, ZnO_2, CuO	<0.65 (as less as possible)

*means should be not less than 0.65%

To produce one ton of limestone (LD grade having less than 1.5% silica content), its producer has to pay Rs. 90/- per ton as royalty to the Government of India (vide gazette notification dated 1 September 2014; G.S.R.630(E), Ministry of Mines, New Delhi). In all other cases, the rate of royalty is Rs. 80/- per ton. Royalty rates vary when limestone is used either as a minor mineral (lime burning, making calcium carbide as a cottage industry, etc.) or a major mineral (cement, steel, fertilizers, etc.). Similarly, royalty rates of dolomite and marl are Rs. 75/- per ton and Rs. 60/- per ton, respectively. These rates are fixed and altered periodically in exercise of the powers conferred under subsection (3) of Section 9 of the Mines and Minerals (Development and Regulation) Act, 1957, abbreviated as MMDR-1957 and as amended in 2015 [11].

1.3 Limestone Versus Dolomite: Mineralogical and Chemical Composition

Limestone refers to a group of sedimentary rock predominantly composed of calcium carbonate ($CaCO_3$). The $CaCO_3$ content in limestone may go up to 99% with some content of magnesium carbonate ($MgCo_3$) or other impurities. Mineralogical limestone is composed of calcite ($CaO = 56\% + CO_2 = 44\%$) and in recent deposits also called aragonite. Limestone, when pure, contains 90–99% of $CaCO_3$ and is classified as high calcium limestone. When magnesia or MgO percentage is 18% or more, it is categorized as magnesian limestone. Limestone with magnesia in the range of 05–08% is described as low magnesian limestone. Other classifications of limestone are—massive, flaggy, marly, milliolitic, variegated, blocky, karsty, and argillaceous limestone. Theoretically, dolomite contains a $CaCO_3$:$MgCO_3$ ratio as 54.35%:45.65%. Thus, the mineralogical and chemical compositions of these carbonate rocks determine their suitability for the specific requirement of industry in which they are consumed or used.

The predominance of calcite in rocks forms the basis for the categorization as limestone or dolomite [$CaMg(CO_3)_2$]. Limestone is also referred to as high calcium limestone, dolomitic limestone, or calcitic limestone (see Fig. 1.1). All of these carbonate rock forms are designated with the change in percentage carbonate and magnesium (Table 1.7). For the purpose of mineral categorization, dolomite is covered under the minor mineral category and falls under the jurisdiction of the individual state governments.

In terms of chemical composition, dolomites resemble limestone except that MgO is a large and important constituent. However, dolomitic and calcitic limestones are readily differentiated by treatment with Lambert's solution or with hydrochloric acid. Calcitic limestone turns blue on treatment with Lambert's solution, whereas dolomite does not change colour. Dolomite rocks are more dense and less soluble in acids than ordinary limestone.

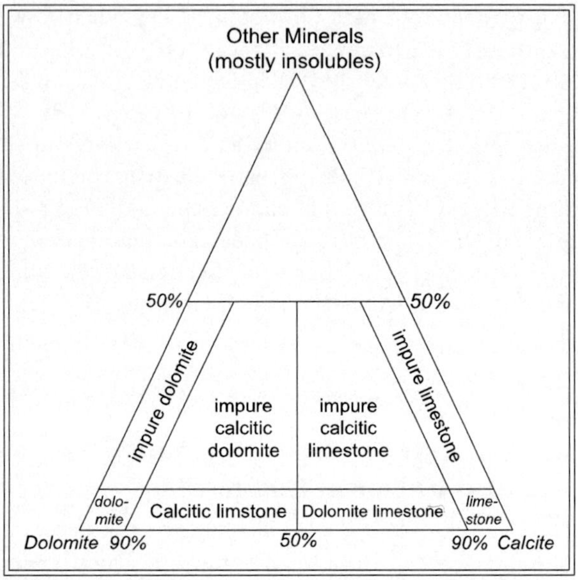

Fig. 1.1 Mineralogical classification of carbonate rocks. *Source* Carr and Rooney [6]

Table 1.7 Mineralogical classification of carbonate rocks

Type	$MgCO_3$ approximate equivalent percentage	MgO approximate equivalent percentage
	All values in percentage (%)	
High calcium limestone	Up to 2.3	0–2.1
Magnesium limestone	2.3–4.4	>18
Low magnesium limestone	–	5–8
Dolomitic limestone	4.4–22.7%	2.1–10.8
Calcitic dolomite	22.8–41.0	10.8–19.5
Dolomite (dolostone)	41–45.4	19.5–21.6

Limestone and dolomite contain certain impurities that may be classified as homogenous or heterogeneous. *Homogenous* impurities are introduced at the time of geological deposition and remain well dispersed throughout the formations as an integral part of the rock. *Heterogeneous* impurities are depositional later and are found in places like crevices, bedding planes, etc. Homogenous impurities are generally present in the form of clay, silt, sand, shale, chert, flint, clastic-quartz, etc. Depending upon the nature of impurities, limestones and dolomites are described with descriptive prefixes, such as shaly, cherty, clayey, siliceous, argillaceous, and ferruginous.

1.3.1 Role of Silica in Limestone and Physico-Mechanical Properties

The reactivity of different forms of silica in limestone with calcium oxides is important. It has been observed and considered that impurities from such reactions should always be investigated in order to classify limestone as cement grade. Besides silica, other impurities, such as alumina and oxides of iron, can also be important for a particular industrial process and shall be kept as per the requirement.

Physico-mechanical properties of limestone provide other essential information from mining and excavation angles. The hardness, moisture, coarseness, crushability and grindability indices, and compressive strength of limestone have direct relevance with the unit operations of mining and limestone production processes [5]. The possibility of reducing the limestone to the required fineness will depend on their hardness and strength parameters, which, in turn, depend on the mineral composition and microstructure of the rock. Regardless of the practices followed in different countries, the general experience indicates that the limestone for production of general purpose cements (OPC, PPC, etc.) is grinded to about 90 μm, and in the production of high strength cement, a fineness of about 45 μm is essential.

Apart from this, machine mining and deployment of continuous miner and surface miner in mining production processes also require these properties. Uniaxial compressive strength of limestone ranges from 10 to 300 MPa, and similarly, the hardness of limestone on Mho's scale varies from 2 to 5 (medium range). The hardness variability is directly connected with the grain size, nature of cementing material, porosity, degree of weathering, natural moisture content, etc. Some other related limestone properties are as follow:

1. The plasticity of limestone increases with reduction in grain size.
2. Hardness decreases with increase in porosity and degree of weathering.
3. Hardness decreases with water content in limestone.
   (This means limestone rocks with high moisture content are comparatively less hard.)
4. Hardness increases with increase of carbonate percentage in limestone, e.g. clay, calcareous clay, and clay-marl.

Thus, it is apparent that the compressive strength and natural moisture content of limestones, apart from their effect on the hardness, have direct relevance to their use in cement manufacture. For the dry process of cement making, the natural moisture content of carbonate rocks is generally restricted to 5%, and for natural marls, it is generally restricted to 10%.

1.4 Geological Occurrences in India

Limestone deposits of different varieties occur in several states of India from north to south and from east to west. They occur in ample quantity, but the distribution is not uniform. To describe them in the geographical and geological time scales in which they are formed (stratigraphical sequences) have been considered, and accordingly, the following subsections are given here.

1.4.1 Limestone Belts of India

Archaean, Vindhyan, and Tertiary limestone deposits are found in the states of Chhattisgarh, Rajasthan, Andhra Pradesh, Karnataka, and Madhya Pradesh in large quantities. These extensive deposits serve as the main feed for the number of cement factories located nearby. A map showing important geological horizons containing limestone deposits is given in Fig. 1.2.

Most of these deposits are of good quality cement grade and are fit for many industrial uses. Archaean deposits are the oldest, and recent deposits are youngest. *Limeshell* and *Kankar* are also good raw materials for hydrated lime preparation and cement manufacturing. Similarly, soft marl and miliolitic limestone, when used in klinker preparation with a sweetener, fulfil the requirement of limestone as a raw material. Both simple deposits and complex deposits are being exploited. Thus, the future of limestone use in industry, including its excavation from mines, is very very good and bright. A summarized table derived from an authentic source has been prepared to show the locality in which these deposits are available (Annexure A). Readers can get an idea of the exhaustiveness of limestone reserves from it.

Limestone rocks are originally sedimentary in nature and have been subjected to various degrees of metamorphism as a result of which silica and clay minerals present in the original rocks have been converted into various calc-silicate phases, depending on the type and extent of metamorphism. Thus, in addition to carbonate quartz, graphite, pyrite, etc., also may be present in limestone rock, and due to these minerals, the texture and mineralogy change [7]. From geological studies, it is made clear that crystalline limestone deposits, dolomitic limestone deposits; high magnesium content limestone deposits; chemical grade, steel grade, or blast furnace (BF) grade limestone, etc., are all minable (i.e. they have not posed any problems in mining so far). However, their end industrial uses need to be moderated as per the industrial needs and requirements.

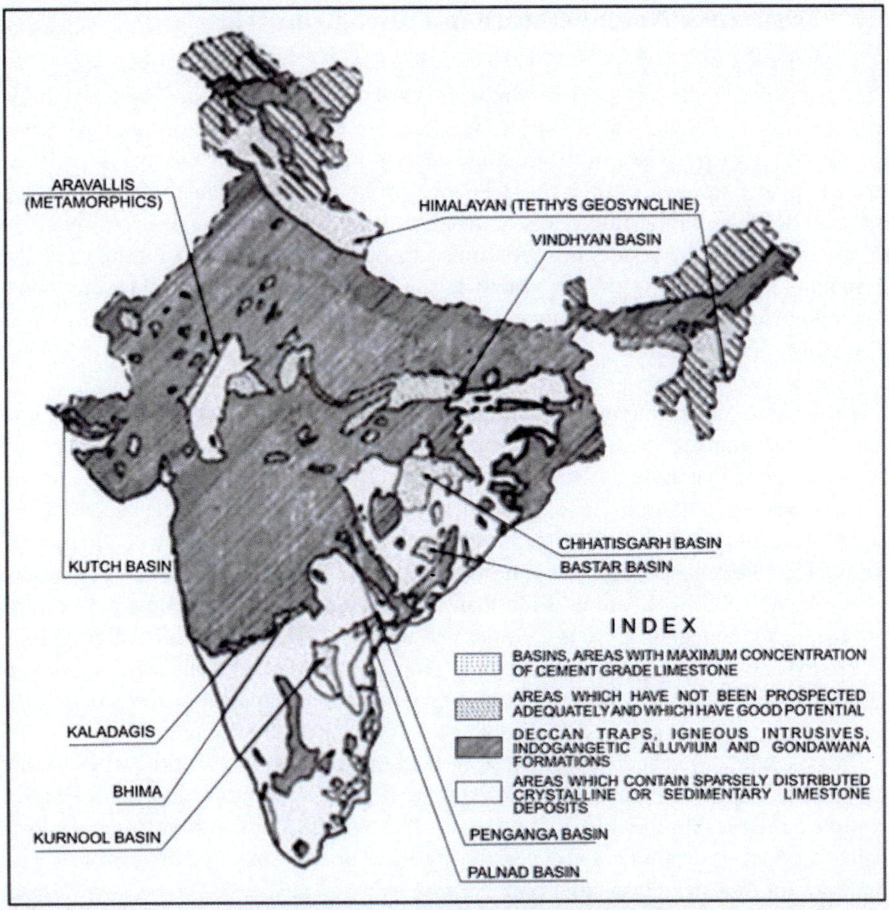

Fig. 1.2 Map of India showing geological horizons containing limestone deposits. *Source* NCCBM [7]

1.4.2 Mainland and Coastland Deposits

In India, geological occurrences of limestone are reported from both its mainland and coastland. Limestone and dolomite both are of sedimentary origin, meaning they have been deposited at the sea bottom as chemical precipitates due to changes in temperature and the chemical composition of seawater [8]. According to the geological age of formations, they are also classified as *Archaean*, *Vindhyan*, *Tertiary*, *Jurassic*, *Carboniferous*, *Cretaceous*, and Meso- or Neo-Proterozoic limestone. Marly and miliolitic limestone are found near the seacoast, and karsty limestone is mined at some places of north-eastern India, where different geological ages of rock have been reported. Seabeds (marine topography) and lake beds also contain significant limestone and carbonate rock.

1.5 Limestone Nomenclature in Other Countries

Limestone nomenclature in other countries is not the same as that designated in India (e.g. in United States, where limestone is covered under the name *crushed stone* for all practical purposes and literature descriptions). In other countries, crushed stone is a term applied to rock that has been broken and/or crushed after quarrying into smaller or irregular fragments or ground to specified particle sizes. About 75% of the crushed stone production continues to be limestone and dolomite, and the remainder is followed by granite, trap rock, sandstone, quartzite, miscellaneous stone, calcareous marl, marble, volcanic cinder, scoria, shell, and slate in order of volume. The crushed stone industry is a major contributor to the infrastructure development and economic well-being.

Limestone and dolomite formations are widespread in the United States with most of the deposits being from the *Paleozoic* age with a few of *Mesozoic* and *Tertiary* age. Limestone as crushed stone is produced by nearly 940 companies at 1959 operations with more than 2000 quarries in 47 states (except Alaska, Montana, and North Dakota). Leading states in order of tonnage are Texas, Florida, Missouri, Kentucky, and Pennsylvania, accounting for nearly 37% of the total limestone output in the United States. Leading American producers, in order of limestone volume produced, are Vulcan Materials Company, Beazer USA/Hanson, CSR America Inc., Martin Marietta Aggregates, and Rogers Group, Inc. Crushed stone mining companies, handling the total limestone output of United States, are in organized sectors, which depicts an altogether different scenario compared to mining in India.

In most countries, where ever limestone is found, this natural resource is accessible for a wide range of industries from construction to agriculture and from easy to complex chemical and industrial processes. Despite the fact that limestone is a low-value product, its value is characterized by its continued use as a major basic raw material for industrial and other applications.

1.6 Limestone Consumption Factor (LCF)

For production of cement and from the view point of rationalization of limestone consumption, the limestone consumption factor (LCF) is used in the cement industry throughout the world with differing connotations. Besides limestone consumption, LCF is also useful in estimating the royalty payable to the state for the limestone that has been mined as well as used for internal material audits of the concerned cement plant. The LCF is determined by knowing the weight and quantity of various raw materials used as ingredients to make clinkers (e.g. limestone, fly ash, slag, or any other additives). The LCF varies from plant to plant and is determined in kilns of cement plants. In India, the LCF usually varies from 1.4 to 1.5. This means about 1.4–1.5 ton of raw material is required to produce one ton of cement. Scientific studies for LCF have been carried out by the National Council of Cement and Building

Materials (Ministry of Commerce and Industry, Government of India), Ballabgarh, Haryana, India, for nearly all of the cement plants in the nation.

1.7 General Characteristics of Limestone Regions

Limestone (calcium carbonate) is a rock as well as mineral (both major and minor). The regions where limestone is found have the following characteristics:

- Groundwater in limestone mines and region is generally hard and basic in nature.
- Land areas containing limestone have copious water below the ground. Such regions contain water in good quantity because limestone and its host rock possess secondary porosity, which transmit water in aquifers where the water table is at shallow depths.
- Caves and sinkholes are formed naturally during the geological formation of limestone.
- There may be a karst type of topography.
  (Karst is a type of landscape/landform/topography, where the dissolution of the bedrocks have created sinkholes, sinking streams, caves, springs, and other characteristic features. Karst landscapes are developed where the bedrock is comprised of an extremely soluble calcium carbonate rocks such as limestone, gypsum, or dolomite. Limestone is the most soluble calcium carbonate rock, and consequently, most karst regions are developed in limestone-rich areas. Thus, the karst landscape and limestone/dolomite are characteristically associated with one another (soluble rock types).
- Areas are identified by ecological indicator plants that grow on the surface.

In India, limestone regions are abundant. Precedence shows that the limestone regions brought positive and significant socioeconomic development benefiting the surrounding population. By and large, these industries, either local or regional, had improved the overall quality of life of that region in terms of literacy, general infrastructure, and basic amenities.

1.8 Limestone in Future Perspective

Within the Indian context of exploration and conservation, limestone offers propitious deal compared to other mineral types. In the future, the limestone exploitation is bound to accelerate and grow faster and reach new heights. Looking at a glance on the environmental impacts, limestone mining (including its transportation during and after mining) is going to be far easier than in the present because of technological prowess. Feasibility or viability of complex limestone deposits for mining in some notable areas (i.e. depth from surface, stripping ratio, homogeneity/heterogeneity, conservation, low MgO content issue) will be possibly explored in the near future.

New industrial applications of limestone also are going to be ascertained and addressed. Precisely, the focus of the Indian government on boosting the nation's infrastructure and rural economy will offer a better scope for limestone mining and its supportive ancillary industries. The current giant Indian cement industry, with an installed capacity of approximately 500 million tonnes, is not only the largest limestone consumer in India but also a big captive mining sector that is occupied mostly by private partners with sizable investments. It is predicted that cement plant numbers and limestone demand will multiply all over the world. Incorporating the concepts of recycling (reuse) and reducing waste and adopting the philosophy of improving energy efficiency, limestone mining has very bright future.

References

1. IBM (2018) Status of reconnaissance permits, prospecting licenses and mining leases in India, Indian mineral year book—2018 (Part-I : General review), Ministry of Mines (Govt. of India), Indian Bureau of Mines (IBM), Nagpur, March, pp 3–1 to 3–6
2. Mukhopadhyay D. (2012); Indian cement industry: a technology perspective, India science and technology report—2010, May, Published by CSIR—National Institute of Science, Technology and Development (NISTAD), New Delhi, pp. T3–146 to T3–157. From web accessed on 12 June 2019
3. MGMI (2006) Limestone for steel making, Mining Geological and Metallurgical Institute of India (MGMI), MGMI transactions, vol 103, No 1 & 2, Apr 2006–Mar 2007, pp 125–136. (Technical paper by G Prabhulingaiah, MS Shekhawat, DS Acharya, AK Jaiswal, R Shankar, presented at MGMI paper meet on 28/03/2017 at MGMI Kolkata)
4. CSIR-CIMFR (2010) Applicability of surface miner in Nongtrai limestone mine of LUMPL in Meghalaya, India—CIMFR technical report of CSIR-Central Institute of Mining and Fuel Research (CSIR-CIMFR), July, p 67
5. Chatterjee AK (2018) Cement production technology: principles and practice. CRC Press, Taylor & Francis Group, p 440. ISBN-13: 978-1-138-57066-5
6. Carr DD, Rooney LF (1983) Limestone and dolomite. In: Lenfond SJ (ed) Industrial minerals and rocks, (5th edn), Society of Mining Engineers of the American Institute of Mining, Metallurgical and Petroleum Engineers Inc., New York IN; Ramaih NRP (2003) Limestone and dolomite resources of Karnataka, geological society of India, Bangalore, p 322. ISBN 81-85867-56-9
7. NCCBM (1985) Comprehensive appraisal of cement grade limestone deposits of India, July, SP 15-85, National Council for Cement and Building Materials (NCCBM), New Delhi, p 160
8. IBM (1982) Monograph on limestone and dolomite, publication cell, technical consultancy, mining research and publication division, Indian Bureau of Mines (IBM), Nagpur, Dec, Chapter 02 (pp 2–1 to 2–16) and Chapter 05 (pp 5–98)
9. IBM (2017) Indian minerals yearbook, Vol III: reviews on minerals, Indian Bureau of Mines (IBM), Government of India, Nagpur, p 3–18
10. https://www.cmaindia.org/key-areas/mines-minerals/. Last accessed on 28 Jan 2021
11. https://ibm.gov.in/writereaddata/files/04192017182323MCDR_2017.pdf

Chapter 2
Excavation of Limestone

Limestone is the main and essential raw material for number of industries, which include cement, iron and steel, fertilizer, chemical, etc. The cement industry is the largest consumer (over 80% of its production) of limestone. The ample availability of limestone in India together with its economical extraction makes this mineral attractive for its mining. Further, the importance of limestone as raw material feed for industrial applications in India can be judged from the following facts:

- India is ranked as the fourth largest cement producing country in the world.
- India is a leading steel producer in Asia, meeting its huge in-house steel requirement through indigenous production.
- Chemical, sugar, and fertilizer industries are large industries in India. These are anchored in the agro-based economy of India.
- The Indian mining industry is fully grown, mature, and has attained a good safety track record for producing minerals in bulk.
- Limestone excavation, although capital intensive, is largely economical and profitable for industrial production.

With this background, we shall discuss and describe different aspects related to limestone excavation and its mining details.

2.1 Basic Concepts

Limestone is found near the surface (called a *surficial deposit*); hence, it is produced mainly from surface mines. By and large, limestone is mined by two methods: *open-cast* stripping for hilly deposits and *open-pit* mining for plain deposits. To meet the industrial requirements of a large country like India, medium- and small-sized mines of various types and different scales of mechanization are operated. Limestone deposits in hills as well as in plain areas are generally covered with little or no overburden coverage. In hilly areas, an entire hill often is composed of limestone

A. K. Soni and P. Nema, *Limestone Mining in India*, Materials Horizons: From Nature to Nanomaterials, https://doi.org/10.1007/978-981-16-3560-1_2

with scant vegetation on the surface, whereas in plains, deposits are exposed on the surface outcropping.

Limestone, when excavated through an opencast method, requires drilling, blasting, loading, and hauling operations. In a conventional semi-mechanized limestone mine, using a shovel–dumper combination, scientific planning of mine, geological reserves, and uncertainties together with environmental protection are the basic issues that need to be conceptualized and addressed. In India, the Mineral Conservation and Development Rules-1988 (MCDR 1988) must be adhered to for limestone conservation (www.ibm.gov.in).

Limestone quarries in India are generally equipped with electric or hydraulic operated shovels, drilling rigs, rock breakers, and tyre-mounted hauling equipment such as dumpers and tippers. Crusher, graders, dozers, and reclaimers are also extensively used equipment. All of these are deployed for the economic advantages of mining to achieve the desired production for a reasonable price and providing a profit.

The cost of moving limestone from quarry to the consumption point often equals or exceeds its sale price, being mineral of low value. Because of the high cost of transportation and the bulk material handling, limestone is usually marketed at short distances and consumed locally (e.g. in a captive cement plant). Thus, limestone handling and its transportation in bulk is a key area, wherein the production cost of limestone can be optimized. Limestone mining can be made holistic with *green mining* and other environmentally friendly concepts and approaches.

In this chapter, limestone exploitation in its entirety, involving excavation methods (mining), safety, clearances, and hot spots together with critiques or gaps in mining and environmental practices, has been covered, which one may consider as a part of mining technology.

2.2 Mining Limestone Using Surface Mining Methods

Achieving bulk mineral production is not possible without a suitable mining method. It is well known that surface mining accounts for the majority of mineral production world over. In India, limestone, which is a low-value mineral found close to the surface, is mainly extracted by opencast mines only. Thus, broadly speaking, surface mining methods are the only method for limestone excavation in India, and the possibility of limestone extraction by underground mining methods in the near future should be struck off (see Sect. 2.3). Two surface mining methods, namely the conventional method and the machine method, are practised in India for limestone production. Both of them will be described in this section.

A method of mining is generally selected and designed considering the scale of operation (production), nature of the rock, topography, thickness of the deposit, overburden, blending required to obtain appropriate grades and other techno-economic parameters (e.g. safety, mineral conservation, and percentage of recovery). The cost of run-of-mine (ROM) production and ease in achieving targeted production (or overall mine productivity) are the key parameters.

2.2.1 Excavation Methods

The following are four applicable methods of the surface mining of limestone.

1. Conventional method
2. Surface and continuous miners
3. Ripping and dozing
4. Rock breaking by breaker.

Conventional methods of mining include drilling, blasting, loading, and transportation. The extraction of limestone is planned, adhering to the methodology mentioned in the mining plan approved by the Indian Bureau of Mines, which is a statutory body of the Ministry of Mines of the Government of India. After the rock is loosened by a drilling and blasting unit operation, the blasted 'muck' is loaded and transported up to the crusher plant using either a dumper–shovel combination, dozer–dumper combination, or shovel–truck (tipper) system. The sized limestone from the crusher is then dispatched to the consumer by various means. At the end of the mine life, a *reclamation by afforestation* or other land rehabilitation techniques as proposed in the *closure plan* of the mine is accomplished. Backfilling operations are needed (in open-pit mining) for restoration of the topography. Thus, the altered topography of the area that gets modified due to the mining activity is brought back to nearly its original shape. The conventional mining methods could be manual, semi-mechanized, or fully mechanized as the case may be.

From among the available methods, the conventional method is the best option for extraction in a plain or hill deposit where ripping and dozing and rock breaking using a hydraulic breaker have limited application. 'Ripping and dozing' and 'rock breakers' are two alternate methods of excavation that reduce hazardous blasting operations. These two methods are further elaborated in Sect. 3.1, which discusses feasible cleaner technologies as options for Indian limestone mines to yield the planned production.

Production-related factors when further analysed and correlated with machinery selection indicate that *surface miners* could be an alternative to the conventional method. Therefore, considering the equipment design parameters critically and its suitability for an area, a detailed assessment and analysis concerning the mining area can be completed. Some theoretical details about *surface miners* are described next.

2.2.2 Surface Miners

A surface miner is a continuous mining equipment used to extract, crush, and load material from the ground surface at once. This machine, if applicable, maybe an alternative to the conventional method of mining. It eliminates drilling and blasting as well as a primary and secondary crushing in the mining of minerals or rock deposits. Surface miners can even replace the overburden removal by bulldozers equipped with rippers that require regular undercarriage servicing and often generate inconsistent end products. Using surface miners allows the driver to effectively bring the primary crusher up to the face of a hill. Thus, surface miners also serve as *in-pit crusher* in another form, allowing mining companies to get rid of primary crushing equipment.

The surface miner concept is not new. It originated from civil engineering applications. Earlier, this mechanical equipment was more or less limited to soil removal, soft land cover digging, soft overburden removal, etc. Later on, its application in mining was extended for surface mining in the removal of softer minerals and materials, such as lignite, coal, sand, and clay. In the late 1970s and early 1980s, the first improved cutting systems were introduced in the surface mining industry that were able to cut harder materials too. Different European companies tried to develop machines based on their earlier experiences in the mechanical cutting of rocks. Some noted companies that have developed surface miners for mining and other similar application areas are

- *Caterpillar*
- *Dosco*
- Voest Alpine
- Thyssen-Krupp
- Wirtgen GmbH
- Huron and Morrison Knudsen
- Man Takraf
- Larsen and Toubro.

These manufacturers developed surface miners from their experience and background for underground mining, civil construction work, and tunnelling. In India, Wirtgen GmbH introduced the surface miners' technology for cutting hard rocks. Later, Easi-Miner® developed and brought in to market a type of continuous surface miners working on the same rock-cutting principles. In recent years, surface miners have been applied and used for mining limestone, marl, coal, lignite, gypsum, asphalt, shale, phosphate, salt, bauxite, etc. Thus, cutting abrasive and hard materials were made possible using surface mining technology with considerable success all over the globe.

Surface miners can be categorized into three types based on their design. These are manufactured by noted heavy earth-moving companies located in Europe, Germany, and the United States (Voest Alpine, Wirtgen, Takraf, Caterpillar, etc.). Depending on the cutting mechanism and position of the cutting drum in the machine, they are categorized in different types as given here.

(a) Surface miners with a middle drum configuration (Fig. 2.1) are the most common in the market and are developed by modifying the levellers used in road construction, including paving and trench cutting.

(b) Machines with front boom cutting drum (Fig. 2.2): Tesmec, Vermeer, and Voest Alpine surface miners are examples of this type.

(c) Some surface miners have a front cutting wheel (Fig. 2.3). KSM type machines from Krupp Fördertechnik and TSM machines of Tenova TAKRAF belong to

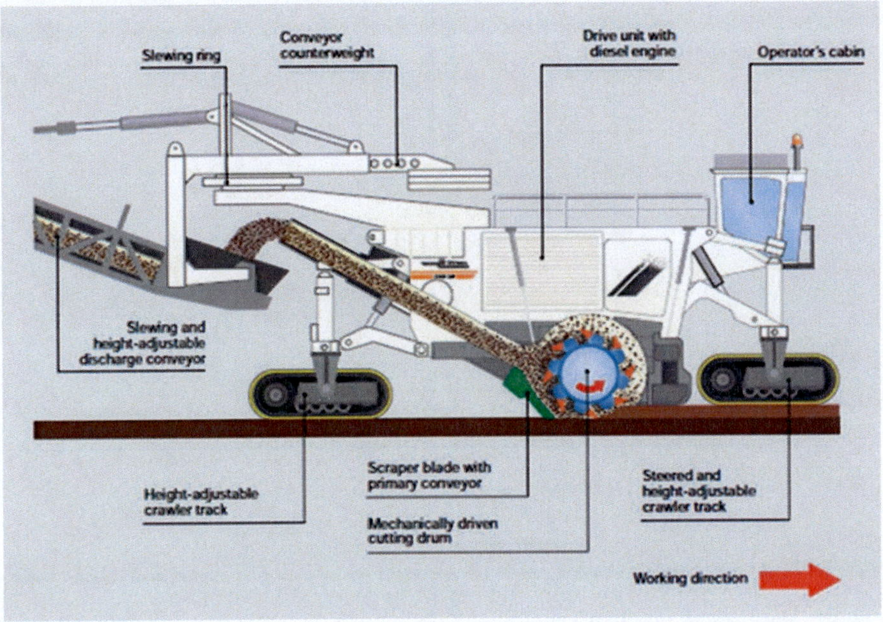

Fig. 2.1 Machines with middle drum configuration

Fig. 2.2 Machines with front boom cutting drum

Fig. 2.3 Machines with the front cutting wheel

this category. KSM machines are three times bigger than the largest surface miner machines and consequently are three times heavier with a proportionately high installed power. KSM machines are specially designed for large-scale operations in coal seams and overburden. Tenova TAKRAF machines are used in coal and iron ore mines. A brief comparison of basic technical parameters of the three types of surface miners is given in Table 2.1.

Some of the different types of surface miners discussed earlier have not been tried in India. Their technical specifications keep on upgrading with time and with the advancement of the engineering knowledge base—be it mechanical, electrical,

Table 2.1 Basic technical parameters of surface miners

Parameters	Type of surface miner		
	Middle drum	Front cutting boom	Front cutting wheel
Cutting width (mm)	250–4200	5250	7100
Cutting depth/height (mm)	0–800	1000–5500	0–2900
Capacity	For all machines, the output is related to material characteristics		
Weight (t)	40–190	135	540
Installed power (kW)	450–1200	750	up to 3340
Manufacturers	Wirtgen, Bitelli, L&T and Huron	Vermeer, Tesmec, Voest Alpine	Krupp Fördertechnik & Tenova TAKRAF

Table 2.2 Surface miners in limestone mining applications in India

Make	Model no	Name of company where deployed	Mineral	Year	Remarks
Wirtgen	1900 SM	Gujarat Ambuja Cements Limited	Limestone	1994	First demo version in Himachal Pradesh nit; (middle drum type) Now deployed at Kodinar unit, Gujarat
Wirtgen	2100 SM 2500 SM	Madras Cements	Limestone		Pannedam limestone mine
Wirtgen	2100 SM	UltraTech Cements	Limestone (marly)		Kovaya limestone mine of GCW
Wirtgen	2200 SM	Madras Cements	Limestone		–

GCW Gujarat cement works

or electronic. In India, the surface miners used in the mining industry were first introduced in limestone mining by Wirtgen GmbH of Germany, although surface miners had been operational in other parts of the world since 1983. The literature indicates that surface miners were first introduced in a gypsum mine in South Africa. Surface miner applications in India (Table 2.2) and the statistics indicate that nearly 18 surface miners were operational in limestone extraction in 2010 with various degrees of success rate.

The application of surface miners has been successful in India for the mining and sizing of soft- to medium-hard limestone only in Gujarat and Tamil Nadu states. In comparison with the limestone deposits in the northern, central, eastern, and north-eastern states of India, formations in Gujarat and Tamil Nadu are softer, so they easily can be mined with surface miners. It may be noted that all of the surface miners running in Indian mines are machines with middle drum configuration, and the majority of them are different models of Wirtgen miners. The statistics for the deployment of surface miners to extract other minerals indicate that a total of 105 were deployed in India and of these 40 were used in the coal sector.

The advantages of surface miners are numerous. These machines are best suited for '*selective Mining*', meaning being deployed for selective extraction where it is essential to extract minerals from partings, overburden, concentrated/low-grade ore veins or seam/beds. Some of the advantages are listed as follows:

- Provides clean working conditions (e.g. stable surfaces, high walls, and working faces)
- Elimination of hazardous blasting operation, providing an alternative to conventional drilling and blasting technique
- Elimination of primary crushing and sometimes secondary crushing as well
- Reduction of nuisances (e.g. ground vibration, noise, and dust)
- Higher productivity because of continuous cutting and crushing operations
- Beneficial applications in projects with sensitive surroundings.

Surface miner applications initially were developed for soil, soft rocks, and loose rock material. The uses were further extended for harder material and mineral excavations. Despite the lucrative benefits of continuous surface miner technology and certain successful applications in different parts of the globe for mining of different minerals and ores, surface miners have yet to become universally popular. Boundary conditions for surface miner applications are quite often governed by natural factors that are beyond human control (e.g. geological conditions, topography, and severe climatic factors). The rock mass, material properties, and characteristics (e.g. strength indices, elastic properties, density, brittleness, abrasiveness, moisture content, joint concentration and disposition, and stickiness) are some geological parameters. Machine configurations for related factors (e.g. cutting tool design, drum weight, engine power, and a specific energy for cutting) are considered over the natural factors. Application-specific parameters, such as production target, and mode of operation (i.e. windrowing or conveyor loading), adequacy of space for machine operation, fragmentation desired, dry or wet output, and selective mining are other important factors in the deployment of surface miners at a site.

2.2.3 Open Cast Versus Open Pit and Hill Versus Plain

When excavation for mining limestone extends above the ground, it is called *opencast mining*, and when extended below the ground, it is called *open-pit mining*. The opencast mining of hill deposit formations has further classification in terms of *mine benching* and *stripping,* whereas for open-pit mines, only the benching system is applicable. In Chap. 4, these methods are described through case studies, which one may consider as site examples for different topography and field conditions where the limestone deposits occur naturally.

2.2.4 Mining and Environment Overview

Any mining activity has environmental implications. Therefore, it is imperative that these aspects must be in mind right at the beginning so that mining and the environment are both maintained in harmony. The very purpose of preparing this section is to give a *suggestive checklist* (see Boxes 2.1 and 2.2) that highlights various factors in mining vis-à-vis the environment to provide the needed focus and due consideration. For troubleshooting an industrial problem, the best solution can be arrived at only when the practical subject-specific experience is used. It is well known that experiences gained from different sites and case studies are best for practical solutions to be implemented as per the requirement.

An *overview of limestone mining* can be subdivided into two broad areas of *mining* and *environment* with the presumption that the limestone is mined by surface mining methods only. On examining the checklist topics listed in the boxes, it is found that each of these topics is very vast and covers different parameters as well. Hence, their elaborate description is beyond the scope of this book. The topics mentioned in the boxes here will certainly attract the reader's attention and prompt them to further explore the specifics of a particular subject. This overview skips underground mining and environment-related aspects and covers only surface mine and the mining environment, considering that in India limestone is mined by surface mining methods only.

Box 2.1: Mining

- Environmental-Friendly Alternates of Opencast and Open-Pit Mines
- Project Planning and Scheduling
- Production Planning and Cost Overrun
- Optimization of Various Unit Operations Including Slope Stability Aspects
- Economics of Mining Operations
- Plant, Equipment, and Machinery—Design, Maintenance, and Cost Economics
- Overburden, Interburden, Waste Dump, and its Management
- Human Resource Planning and Management
- Mine Safety—Statutory Requirements and Compliance
- Quality Control of Raw Material Mined.

The environment is the prime concern for any opencast or open-pit mine. Promoting a natural way of environment restoration assumes great importance in mines, where dealing with the mined-out area and dump sites attracts maximum attention. Both an *integrated approach* and *principles for the management of environment* need to be applied in various limestone mines of all sizes as these are more cost-effective and preferable to other methods [12]. For big production operations and large capacity mines, financial restraints are manageable and not of much concern. However, for small mines, reclamation by natural means is the best solution. Research studies previously carried out a point that 'natural reclamation or restoration of derelict land' is a simple and easy way to get the green area back. Methods that are not cost-effective have a chance of only theoretical survival.

Improper biological reclamation of mined-out areas and dump sites in limestone mines causes a lower rate of survival of plants. In the biological reclamation of derelict land, the organic carbon and water holding capacity of the land should be increased first, which in turn supports the growth of vegetation. Biological reclamation is something more than mere planting of trees. It starts with grasses and in later years is followed by shrubs and trees, as this type of succession increases the overall fertility of the soil permanently. The use of fertilizer and chemicals to improve fertility is a temporary measure. As mentioned earlier, an integrated approach and principles for best practice management need to be applied. The following points are noteworthy in this regard.

1. In all plantation work, preference should be given to local species in preference to exotic ones and mixed cultures in preference to single cultures.
2. Some selected species that can grow quickly, stabilize effectively, and improve the soil quality with time—besides being economically useful—can be planted for reclaiming the wasteland generated due to limestone mining.
3. For direct visibility, there can be some cosmetic planting for early results, thereby hiding the scars of damaged land.
4. The use of microorganisms and other organisms living in the soil for enhancing the fertility of the land in a natural way should be adopted.

Box 2.2: Environment

- Pollution—Control and Management in Mines and Mining Areas (Land, Water, Air, and Noise)
- Environmental Impact Assessment (EIA) for Mining Projects. (Physical Impact, Biological Impact, Socio-economic Impact, etc.)
- Environmental Indices (Air, Water, and Noise Quality)
- Environmental Monitoring
- Atmospheric Environment
- Environmental Management Based on an Environmental Management Plan (EMP) [13]
- Reclamation or Revegetation in Degraded Land Areas and Green Belt Development
- Environmental Management System (EMS)
- Environmental Audit
- Mining in Sensitive Areas and Different Ecological Set-ups including Coastal Areas governed by the Coastal Regulation Zone (CRZ) Regulation
- Ecological Indicators Plants
- Carrying Capacity of Mining Regions
- Life Cycle Analysis
- Mine Waste and Biomass Studies
- Human Dimensions and Quality of Life
- Legislation and Indian Environmental Standards
- Financial Incentives for Environmental Protection.

2.3 Underground Mining of Limestone

Minerals and *forests* are two major natural resources that are equally important for human sustenance. However, the dilemma over the use of these resources is often noticed because, if a forest is protected, the mineral cannot be mined, and if the mineral is mined, a forest cannot be saved from destruction. Therefore, the question arises: Can we exploit minerals without disturbing precious forest? Yes, it is possible using underground methods of mineral extraction. Considering this basic principle and assessing the need for mineral, this section on the underground mining of limestone has been described. Wherever mineable reserves of limestone are found in forest areas, instead of extraction using surface mining, applying underground methods (as a policy) will make limestone mining projects economically viable for the mining companies. Through this approach, deep-seated and difficult to extract limestone of different grades can be exploited without affecting the surface environment. With the necessary incentives and further research attempts, an appropriate methodology for the underground exploitation of limestone reserves is possible.

Limestone deposits at great depth or with progressive overburden (i.e. deposits dipping into the ground or hill) are cheaper to extract by underground means compared to surface methods. Economical extraction, low environmental risk, and better conservation of mineral wealth are the key points, making the adoption of underground methods in such cases favourable. The long-term gain overshadows the short-term drawbacks of higher capital investment in underground methods, especially considering the selection of an appropriate mining method. In the Himalaya region, the underground mining of limestone had been experimented for the first time in India, in the year 2000 considering the benefits of going underground and keeping in view the environmental sensitivity of the region [1]. However, underground mining of limestone was found uneconomical at the small-scale level.

As regards the option of underground mining of limestone, a few questions are often asked:

Are such practices in use elsewhere and is it cost-effective? The answer is *yes*. In developed and mineral-rich countries, such as the United States, Canada, Australia, and South Africa, underground mines continue to exist for the commercial production of limestone products.

What are the benefits? Table 2.3 gives a concise comparison of the advantages and disadvantages of selected underground alternatives and surface alternatives. The underground methods mentioned in the table are those that are likely to be applicable for limestone mining. Based on the characteristics of various methods of mining and their theoretical values as given in Table 2.3, one can understand that the mining cost is on the higher side by underground means. This indicates that it is not economical to excavate limestone using underground methods. However, for the co-existence of man, industry, and the environment in an ecologically sensitive area or region with higher fragility, mining using underground options could be a feasible alternate. From a production angle, underground options always stand second compared to surface mining.

Table 2.3 Comparison of selected surface and underground methods

Characteristics	Surface methods			Underground methods (self-supported)		
	Open pit (mechanized)	Manual mining	Open cast (mechanized)	Room and pillar	Shrinkage	Sublevel
1. Mining cost (%)	25	30	25	40	45	50
2. Production rate	Large scale	Small scale	Large scale	Large	Moderate	Large
3. Productivity	High	Low	High	Moderate	Moderate	High
4. Capital investment	Large	Low	Large	Low	Moderate	High
5. Development rate	Rapid	Slow	Rapid	Rapid	Moderate	Moderate
6. Depth capacity	Limited	Limited	Limited	Unlimited	Unlimited	Unlimited
7. Selectivity	Low	Moderate	Low	High	Low	Moderate
8. Recovery	High	Moderate	High	Moderate	Moderate	High
9. Dilution	Moderate	Low	Low	Low	High	Moderate
10. Flexibility	Moderate	High	Moderate	High	Moderate	Low
11. Stability of openings	High	Moderate	High	High	Moderate	Moderate
12. Subsidence	Low	Low	Low	Moderate	High	High
13. Environmental risk	High	High	Very high	Low	Low	Low
14. Waste disposal	Extensive	High	Minor	Minor	Minor	Minor
15. Health and safety	Moderate	Moderate	Moderate	Moderate	Moderate	Moderate
16. Others	Low breakage cost, rainfall and weather problem, large-scale best	High breakage cost, waste intensive and labour dominated	No waste haulage, low breakage cost, large-scale best	Highly mechanized caves with pillar recovery good ventilation	Gravity flow in stope, labour intensive, simple method	Gravity flow in stope, mechanized, large blast, good ventilation

Note Intentionally, the terms used in the table are relative and not absolute

2.4 Critiques and Gaps in Mining and Environmental Practices: Some Facts

Environmental practices in limestone mining have several gaps that can be categorized as lacunas. Critiques and analysis from different field case studies highlighting important issues are given here. Some specific issues will be further detailed later.

- Opencast mine layout
- Long-term and short-term mine planning
- Systematic mining in unorganized sectors
- Small-scale mining and scientific planning
- Appropriate mine closure and fund provisions
- Blasting, ground vibrations, noise, and air overpressure
- Environmental planning and practices
- Air pollution and fugitive dust
- Water pollution, mine water conservation, and hydrological problems
- Land management practices and loss of vegetation
- Disposal of waste
- Socio-economic dimensions of limestone mining
- Lack of efforts for reclamation and ecological rehabilitation of mined-out areas
- Profitable use of the mined-out land area
- Management: corporate governance, statutory compliances, labour management, etc.

2.4.1 Problems and Prospects of Opencast Mine Safety

Safety and productivity go hand in hand. If safety standards are reduced, mining may become unsafe and unproductive. This may lead to the loss of time, money, and lower employee morale as well. For an organization or industrial operation (including mining) to achieve the best, safest, and most productive environment, each organization must be equipped with adequate resources, effective communication, and empowerment.

Opencast mine safety in the limestone mining industry can be enhanced or dealt effectively with knowledge of *safety engineering* and *safety management*, which are the modern and newly emerged tools. It is beyond a doubt that safety is a major concern for the mining industry as a whole and is required for review by all mining professionals. We all want to go back from work safely and without inviting any health problems. Now let us understand or define what these terms exactly mean.

- **Safety Engineering**—*The identification, evaluation, and control of hazards in man–machine systems that contain the potential to cause injury to people or damage to property*

- **Safety Management**—*Consists of a set of safety program elements, policies, and procedures that manage the conduct of safety activity. Safety management provides the solutions within which the techniques of safety engineering are applied.*

Safety engineering and *safety management* are the two sides of the same coin integrated into one. Thus, safety engineering is the physical and mathematical side of injury and damage prevention, and safety management is the administrative or software side of such prevention. Some key points to understand these two aspects are described here.

(i) **Zero Harm**: In India, if we examine the mining companies that had a successful accident and harm prevention programs, we find that they invariably started at the top level of the organization. There are three guiding principles in this context.

1. Known support of the management to lower-level workers
2. Real commitment at every level (with action and not verbal)
3. Individual empowerment (to correct unsafe conditions and practices).

These guiding principles are essential to create a culture of prevention. *Zero harm* can be successfully achieved using an *organized approach* for everybody associated to have effective prevention.

(ii) **Accident Prevention**: Accidents are preventable. Successful accident prevention requires a minimum of four fundamental activities:

1. A study of all working areas to detect and eliminate or control physical hazards that contribute to accidents
2. A study of all operating methods and practices
3. Education, instruction, training, and discipline to minimize human factors that contribute to accidents
4. A thorough investigation of incidences to determine contributing circumstances.

Many persons, either through ignorance or lack of understanding, unfortunately, believe that accidents are the inevitable results of unchangeable circumstances, fate, or a matter of luck. It must be emphasized that accidents do not happen without cause. The identification, isolation, and control of these causes are the underlying principles of all accident prevention techniques.

No person in a supervisory position can be effective in his responsibility for accident prevention unless he fully believes that accidents *can* be prevented and constantly strives to achieve this result. There are several causes of accidents. Here, they are divided into the following three major categories.

(a) Unsafe acts of people
(b) Unsafe physical or mechanical conditions
(c) Factors beyond one's control (unavoidable).

Statistics indicate that 88% of all accidents are caused by unsafe acts of people, 10% by unsafe conditions or mechanical failures, and 2% by factors beyond the control or is unavoidable. Thus, the basic causes for high rates of accidents or injuries are *unsafe acts* and *unsafe conditions* or *both. Unsafe acts* mainly arise due to human error or behaviour-related causes due to ignorance or lack of alertness. *Unsafe conditions* may occur due to insufficient mine design, unanticipated rock behaviour and geology, ill-maintained equipment, or inadequate supervision. Accidents may be caused by a combination of these factors. Improvement or elimination of hazardous conditions is necessary for reducing injuries and improving safety.

(iii) **Risk Management and Assessment**:*Risk is the chance of injury or loss.* This concept includes both the likelihood of a loss and the magnitude. Various approaches to risk management and assessment are prevailing. *Risk probability* and *risk severity approaches* are combined to arrive at a *safety risk matrix*, leading to acceptance or rejection of the practice. Every activity or operation—whether in mining or any other industrial operation—should be split into these categories.

- Elementary activities
- Probability and severity of occurrence (judged or calculated)
- Safety risk (worked out)
- Methodology for safe working (worked out)
- Standard safe procedure (worked out and known as a method statement and a safety management plan).

This entire procedure makes use of special tools, remote operation, etc., to spell out the recommended procedures and to be documented in the form of a manual where work persons than are trained in the procedure.

There are alternate methods also available, and choices can be made from them. However, the *continuous improvement approach* (CIA) based on a PLAN–DO–CHECK–ACT (PDCA) strategy is more often adopted to achieve better results. The CIA broadly includes policy, planning and organization, implementation, measurement and evaluation, and a management review model to be done in this sequence.

We all make judgements about risks according to our perceptions and beliefs. Accordingly, *risk evaluation* is one significant component that should be accorded the highest priority in developing responses for risk treatment. Risk evaluation aims to sort the risks into groups like *extreme risk*, *high risk*, *substantial risk*, *moderate risk*, and *low risk.* These are the words that determine the level of management response and effort required.

The risk analysis process generates a set of risk ratings that are used to set priorities. Within the company, risk studies are undertaken in a structured way, making use of information, judgement, and experience from a range of sources, so there will be a degree of confidence in conclusions. Since the perceptions of risk evaluations are those of a group and may not necessarily match with those of a community, there is the possibility that one may meet unexpected resistance to the plans. *Total risk* can be determined quantitatively also using different ratings for probability risk (Rp), a consequence of risk (Rc), and exposure risk (R_E). Total risk (R_T) is calculated using the following formula.

$$\text{Total Risk}(R_T) = R_P \times R_C \times R_E \tag{2.1}$$

where

R_P Probability of occurrence of risk
R_C Consequence of risk
R_E Exposure level to a risk.

Based on the evaluated risk, *the risk treatment* is made, and a safe environment is created.

(iv) **Safety Management System**: A safety management system broadly includes

(a) Management commitment at senior level
(b) Safety policy
(c) Safety information
(d) Safety culture
(e) Setting safety goals to identify the responsibility and accountability of various personnel
(f) Hazard identification and risk management
(g) Safety reporting system (can be established based on accident statistics collection, analysis, and actions)
(h) Safety audit, monitoring, and assessment
(i) Accident and incident reporting and investigation
(j) Safety orientation and recurrent training
(k) Emergency response plan
(l) Documentation.

Within a safety management system, there lies *safety monitoring*, which has links with risk evaluation too. In general, the safety performance used in mines is primarily monitored by the Directorate General of Mines Safety (DGMS) in India, but at the organizational level, it is monitored through the following:

(a) Standing committees on safety at the India Ministry level with the participation of other educational institutions and research organizations
(b) Internal safety organization at corporate levels consisting of safety boards
(c) Subsidiary-level tripartite safety committees comprising DGMS workmen representatives and management representatives
(d) Safety committee at the area level or mine level
(e) Workmen's inspector.

Good safety motivation together with responsibility and accountability through training the front line of manpower (i.e. workers, supervisors, and managers of the actual workplace) in the mine is needed in opencast mine safety.

2.5 Statutory Compliance and Clearances Concerning Limestone Mining

In India, compliance with the law is an obligation of every mine operator. Most of us would agree that human beings are frequent violators of the 'rules' wherever they might be. Assuming that the obeyance of rules, meaning safe operating procedures, less risk and better output. Thus, a simple equation is evolved:

$$\text{Compliance} + \text{Clearances} = \text{Safe and Productive Mine}$$

By complying with the law alone, one cannot achieve the desired goal of safety and productivity, but its successful implementation gives an element of confidence to every manager/supervisor and employee of the organization.

A partial list of compliance for a day-to-day mining operation is as follows.

1. Consent to operate (CTO)
2. Permission for carrying out blasting in mines under the Metalliferous Mines Regulation (MMR) of 1961 and the Mines Act 1952
3. Permission for storage of explosives to carry out blasting in mines under the Explosive Act
4. Annual statutory compliance for air quality, water quality, pollution abatement, monitoring, etc., in mines
5. Statutory compliance and reporting for heavy earth-moving machinery (HEMM) operations.

Concerning limestone mining in India, a Web-based system called ***PARIVESH*** is the current and existing resource (Sect. 4.5).

2.6 Hot Spots in Limestone Mining

The cement industry is an innovative, forward-looking industry with an ever-increasing demand for limestone production. It is obvious that the increase in limestone production capacity will come from new production points or the expansion of existing plans. Productivity increase includes the reduction in the cost of production, which can be achieved by minimizing 'down time' through modern approaches and innovative technologies in cement-making processes or through the economy in raw material production cost. Such limestone mining areas where cost could be minimized are referred to here as ***hot spots***. Here, our emphasis is to concentrate only on raw material aspects (i.e. limestone production from mines).

Some important *hot spots* that are of concern in this area are

1. Raw material handling using equipment and energy saving
2. Cutting down of mass transport cost for limestone transportation

3. Improvements in unit operation costs of mining to cut down the overall cost
4. Circular economy a new concept/approach.

For the cost reduction areas mentioned, readers are advised to derive the cost reduction modules from available technical literature according to the field-specific conditions. Improvements in the unit operation of mining to reduce the ROM cost are described throughout this book at appropriate section separately [e.g. best mining practices or best practice mining (BMP), routine condition monitoring (RCM) (see Annexure B), etc.].

The benefits of discussing hot spots here are immense, and this can be achieved by introducing the recent technical developments made in opencast mining areas. However, only key grey areas concerning limestone mining are explained here.

2.6.1 Energy Saving and Raw Material Handling

Conserving energy and reducing costs are not new initiatives for limestone mining in general and the mining industry in particular. Since the initial emphasis on energy efficiency in the 1980s and 1990s, mines are still struggling to control energy consumption per unit of production. It is important to note here that investments in the energy sector create jobs, provide improved technologies, and help to grow the national economy. Energy efficiency also helps in growing small ancillary businesses work through funding and financing opportunities and offers good scope.

Energy efficiency improvement programs in mining are widespread and target all aspects of mining. Some of these programs include managing electricity demand, mine drainage, mine ventilation, generating energy from by-products, capturing waste heat, etc. However, comminution and material handling (including loading and hauling) operations have been identified to have the highest potential for energy saving and efficiency improvements. Diesel-operated equipment used in mining and material handling was identified as the operations presenting the greatest energy savings potential [2]. Similarly, improving the energy efficiency of mineral processing (or coal preparation), plants and material handling operations, compared to other processes, confirm this assertion [3]. Economics remains a drive in the decision to include renewable energy projects into mine development.

Today, *energy management* is a key performance indicator for many mines and is routinely reported in *annual sustainability reports* of individual mines or entire corporations. A particular strategy, for the mine's management to achieve its energy management goals, will depend on the mining method and the mine's unique circumstances.

Regarding mine mechanization, quite often a company will make use of heavy equipment for raw material handling in limestone mining. This could be a shovel, a dumper, or a dozer. Such equipment requires massive investment and makes use of heavy power consumption. Therefore, the economics of production would primarily depend on *machine utilization*, which is indicated by the available time. A group

Table 2.4 Case study of machine breakdown in mines* (Study period = 4560 h)

Machinery	Number of breakdowns	Percentage of total number	Percentage of total time lost	Hours
Machine 1	106	30.7	34.1	409.8
Machine 2	89	25.7	23.2	280.5
Machine 3	82	23.7	17.1	204.4
Machine 4	69	19.9	25.5	305.3
Machine 5	346	100	100	1200.0

Source Sarkar [5]
*means for mines other than limestone

of economic experts in the United Kingdom has assessed that improving machine utilization by even 1%, the monetary savings is to the tune of 80 million pounds for the mining industry in that country. Hence, one can very well assess the importance of machine utilization [4]. In many countries, to get the most out of the available equipment and to improve equipment performance (thereby economics of mining), the concept of *routine condition monitoring* (RCM) is being applied successfully.

Machine utilization is poor in the Indian mining industry. Many factors affect the performance of a machine; for example, heavy earth-moving machinery (HEMM) or digging equipment performance will be dependent on operating conditions, operator's practices, equipment characteristics, the mine plan, and the mine design. The reason for low machine utilization also may be many, including frequent breakdowns, failure of power supply, incompatible supports and machines, poor organization, infrastructure facilities, and lack of spare parts. However, among all of these factors, equipment breakdown plays a major role, as is shown in the results of a typical Indian case study (see Table 2.4). Indicated data in Table 2.4 is for the breakdown of an underground longwall method for mining a face. This case study was equipped with a shearer, armoured face conveyor, powered support, and a belt conveyor network. Similar case studies also can be carried out for limestone mines. It is evident from the data of Table 2.4 that elimination of equipment breakdown—even partially—can reduce the total time lost and improve machine utilization substantially.

2.6.2 Circular Economy

Over the past decade, a new concept of the *circular economy* has emerged. In particular, it is suited to all of the sectors, including the mining and allied metals industries. A circular economy is a *cyclic approach* to manufacturing and resource management. It is simply an improved version of the best mining practices and recycling and reuse. Mining as an industry creates lots of waste (e.g. overburden rocks, pumped out mine pit water, and water treatment sludge). All of these wastes can be reused, either within the production chain or repurposed elsewhere. Waste rock, for example, is often used as mine backfill, as landscaping material, or as aggregate in road construction. Sludge from water treatment can be sold commercially for use in paint and pigments as it is high in metallic content, such as iron. Other by-products of the mining sector can be reused for making construction materials (e.g. coarse-grained sandstone can be crushed to make sand; slate waste can be converted into bricks; calcium from limestone can be used as auxiliary raw materials for smelting; *clinker dust* as a by-product from cement plants can be reused (as briquets, granules, or pellets); low-grade limestone can be crushed and reused in agroforestry; calcium carbide manufacturing or chemical treatment of wastewater as a constituent). In brief and on the whole, in a circular economy, the use of raw materials is optimized, and the intelligent reuse of any waste products created during the mining process, manufacturing, and material processing is promoted. In this way and at the end of the mineral's life cycle, the mineral resources are recovered through recycling, justifying the philosophy: *money saved is equal to money earned*. Past precedence shows that from electronic waste—up to 85%—metals are recovered and reused, making them infinitely recyclable.

Therefore, limestone mining operations and their production sites (mines) also have a greater scope for adopting a circular economy approach to business. Through circular economy approaches or by *reuse*, *value addition of waste products* (repurposing), and ***recycling***, part of the total production cost is recovered. Considering the environmental and societal impact of mining operations, mining companies can minimize the negative or ill effects of mining by sharing their best practices and reducing the waste available to them.

2.7 Latest Trends in Limestone Mining

It is often asked: What is the latest trend in limestone mining? The actual process of technologically induced changes in all business and industrial sectors, including mining, is of digital transformation commonly referred to as *digitalization* and *Industry 4.0* [6]. These terms are very broad and vague. They apply to all equipment, machinery, and processes involved in mining. Thus, both the machinery and mining process can be automated and mechanized. Newer and current technologies make necessary and essential changes in their system to keep pace with the latest trend(s) inevitably. To describe this latest evolution and revolution, the following points are mentioned here.

- Artificial intelligence (AI)
- IT infrastructure in mines
- Internet of things (different apps for compatibility with the Android™ system)
- Automation and robotics
- Advanced simulation and virtual reality (VR) mines [7]
- Data analytics: big data and real-time data tools
- Machine learning for mining applications

- 3D printing technology for mining applications
- Advanced process control: The concept of process-controlled opencast mines and intelligent mines [The mines referred here have a computerized system for the compilation of geological exploration data; processing for deposit modelling and excavation scheduling, such as satellite supported mine surveying and mass calculation; and a GPS-based system for mining operation (e.g. CES system of FLSmidth, GeoPlan®, CADE system) [14].

The increasing depth of open-cut mines has to lead to global positioning system (GPS) inaccuracies. However, recent developments of terrestrial high-accuracy positioning systems, together with the global navigation satellite system (GNSS), can now deliver the accuracy required for open-cut mine machine automation applications [8].

- Central control and dispatch network (where limestone transportation using a *truck dispatching system* with GIS- and GPS-based technology is the current trend in limestone mining)
- Computer-aided evaluation of mineral and ore deposits using *geostatistics* (i.e. kriging or block modelling)
- Computer-aided mine planning, preparation of maps, and inventory, including the selection of HEMM for mines
- Quarry schedule optimization (QSO) and mine production scheduling for ROM production at different locations in a mine
- Continuous surface miners
- Improved rippability factoring for ripping and dozing (through and R&D approach)
- Use of microprocessor in HEMM in mines for remote operation (where infrared-reflective probing sensors are used for night operations in CSM to detect the difference between country rock and minerals, automatic levelling devices to maintain preset cutting depth in CSM, and grade and slope control monitors for mining of dipping beds)
- Computer-aided loading of excavators
- *Controlled blasting* and use of modern tools, including bottom hole initiation techniques, image analysis software system to measure fragmentation (FRAGALYST), blast casting technique, site-specific blast design for improved blast productivity, and a programmable sequential blasting machine (SQM)
- In-pit crusher for limestone crushing near the production point
- High-angled belt conveying where conveying uphill up to the gradient of 80–90 is possible these days. Deploying *snake conveyors*, *sandwich belt conveyor*, and *flexible face conveyors* can be used for open-pit, cross-pit, and pit walls.
- A moving-bed chain elevating conveyor is a trough-shaped conveyor belt that runs in parallel with a chain conveyor. Paddles shaped to the profile of the belt trough are driven by a chain and propel the material and underlying conveyor belt up and down the incline. Only the chain is driven, movement of material and belt being achieved by friction between the paddles and the belt. This system can handle lumpy and wet material.

- Development of a single machine that combines the operation of both drilling and excavating
- Noise-efficient high-capacity compressors (14 bars and 24 bars)
- Ore chute or ore pass for ROM transportation in hilly deposits [9]
- Use of *mobile rock breakers* for secondary blasting
- Circular economy and routine condition monitoring (RCM) for different applications in mining (see Sect. 2.6)
- Utilizing limestone with higher levels of magnesium for cement making (at the R&D stage).

In addition, Australasian Mining and Metallurgical Operating Practices (AMMOP), which describes the significant mining and processing operations in Australia, New Zealand, and Papua New Guinea [10], is extremely relevant for the latest trends. For limestone mining in India, different issues and key areas of mining and the environment can be closely dealt with using the described knowledge of AMMOP. This includes R&D approaches and the latest trends and their implementation, which are possible for use at various levels in operational mines.

2.8 Quality Control in Limestone Mining

After excavation from mines, limestone sometimes requires a proper chemical composition or grade. This grade maintenance is achieved by adding an equivalent raw material that is referred to as *sweetener*. This whole process is what is commonly referred as *quality control and beneficiation*. Normally in steel, chemical, or cement industries, different grades of limestone are used as a raw material feed (e.g. in cement making, the main requirement is that the magnesia and sulphur content of the limestone should be low and the CaO or $CaCO_3$ percentage should be as per the required specification of cement. In another example, getting rid of clay present in limestone becomes essential; therefore, for such instances, quality control becomes extremely essential and important.

In the case of captive limestone mines of the cement industry (only if required), quality control is achieved by *hand sorting* (manual means), *blending*, *screening*, *washing*, *wobbling*, or *flotation* methods [11]. Since limestone is a cheap mineral (cost-wise), it is pertinent to choose a less expensive method of quality control. Since the present position regarding limestone reserves in India is quite satisfactory, it is extremely important to choose and apply a method for quality improvement such that the economic aspects are not overlooked.

For quality control, the limestone is used in three different ways: as an aggregate, as calcium carbonate in the cement industry, and to obtain calcium carbide (CaC_2) after burning in the lime industry. The limestone used for cement and lime manufacturing contains clay impurities that can be washed off. These impurities containing mostly clay are called *tailings* and can be stored in ponds. The calcium carbonate industry operates mainly with deposits of a higher grade (>96% $CaCO_3$) and needs to use flotation to separate calcium carbonate from the unwanted minerals (e.g. graphite and mica). Dewatering devices (e.g. thickening and filter presses) are used for limestone processing in this industry. The aggregates sector does not generate waste or tailings. Therefore, there is usually no need for further mineral processing steps.

References

1. Soni AK (2017a) Mining in the Himalayas: an integrated strategy. CRC Press, Taylor & Francis, p 225
2. DOE (2007) Mining industry energy bandwidth study, U.S. Department of Energy (DOE), http://energy.gov/eere/amo/downloads/us-mining-industryenergy-bandwidth-study. Accessed on 14 Apr 15
3. Awuah-Offei K (2016) Energy efficiency in mining: a review with emphasis on the role of operators in loading and hauling operations. J Cleaner Product 117:89–97. https://doi.org/10.1016/j.jclepro.2016.01.035
4. Subramaniam SK, Husin SH, Yusop Y, Hamidon AH (2009) Machine efficiency and man power utilization on production lines. In: Proceedings of 8th WSEAS international conference on electronics, hardware, wireless and optical communications, Feb, World Scientific and Engineering Academy and Society (WSEAS), Wisconsin, USA, pp 70–75. ISBN: 978-960-474-053-6
5. Sarkar SK (1995) Routine condition monitoring in mining. Oxford and IBH Publishing Company Private Limited, New Delhi, p 147
6. Barnewold L (2019) Digital technology trends and their implementation in the mining industry," In: Mueller C et al (ed) Mining goes digital: proceedings of the 39th international symposium on application of computers and operations research in the mineral industry (APCOM-2019), June, Wroclaw, Poland. CRC Press, ISBN 978-0-367-33604-29), pp 9–16
7. Suppes R, Feldmann Y, Abdelrazeq A, Daling L (2019) Virtual reality mine: a vision for digitalised mining engineering education. In: Mueller C et al (ed) Mining goes digital: proceedings of the 39th international symposium on application of computers and operations research in the mineral industry' (APCOM 2019), June, Wroclaw, Poland. CRC Press, pp 17–24. ISBN 978-0-367-33604-2
8. Rizos C, Lilly B, Robertson C, Gambale N (2011) Open cut mine machinery automation—going beyond global navigation satellite system with Locata. In: Saydam S (ed) Proceedings of second international future mining conference, The Australasian Institute of Mining and Metallurgy, Melbourne, pp 87–94
9. Soni AK (1999) Environmentally friendly transportation of limestone for cement production. Int J Bulk Solid Handling 19(3):329–336. Germany. ISSN: 01739980
10. Dight PM, Douglas B, Henley K, Lumley G, McAree PR, Miller D, Saydam S, Topal E, Wesseloo J, Williams DJ (2013) Mawby M (ed) Developments in open cut mining. In: Australasian mining and metallurgical operating practices, vol 1, Ch. 3, 3rd edn, The Australasian Institute of Mining and Metallurgy (AusImm), pp 47–79
11. IBM (1979) Working of captive limestone mines of cement plant in India, Bulletin No 08, Technical Consultancy, Mining Research and Publication Division, Indian Bureau of Mines (IBM), Ministry of Steel, Mines, and Coal, Govt. of India, Nagpur, Dec, pp 100–109
12. Neri AC, Sánchez LE (2010) A procedure to evaluate environmental rehabilitation in limestone quarries. J Environ Manage 91(11):2225–2237. https://doi.org/10.1016/jjenvman.2010.06.005
13. Saxna NC, Singh G, Ghosh R (2002) Environmental management of mining operation. Scientific Publisher (India) Jodhpur, p 410. ISBN: 81-7233-296-3
14. Soni AK (2017b) Process oriented opencast mine for limestone. In: Proceedings of international conference & expo on mining industry: vision 2030 and beyond, Dec, MEAI Nagpur, pp 188–193

Chapter 3
Limestone Mining, Industry, and Society

Three keywords (i.e., industry, society, and limestone mining) are important to describe this chapter in continuation of the preceding chapter, which explained the mining or excavation process. If we analyse the above three aspects, the relationships that exist between them become clear. A close and in-depth understanding correctly promotes green development and clean excavation of limestone from mines. In order to shape the prosperous industry with a better society (mine and miner), strong policy matters need to be harnessed. In various sections of this chapter, policy formulation, statutory and legal compliances, production cost of minerals, and costs of mining operation—together with contemporary issues, such as corporate social responsibility (CSR) and its role—have been emphasized. To execute or put them into practice, a holistic approach is a must and should be considered for these useful industrial resources.

3.1 Clean Technology Options, Industry, and Society

Changing societal expectations have influenced the way industries involved in the development or extraction of natural mineral resources conduct their operations around the world. Increasingly, communities are demanding more involvement in decision-making around such operations, have expectations of receiving a greater share of the benefits from these operations, and require assurances that the industries involved are appropriately regulated. The combination of increasing pressures on industry performance and the associated societal acceptance of such operations has been described as the *social licence to operate* (SLO). In many ways, the social licence reflects the evolving nature of the relationships between industries, their communities, and other stakeholders. Originally used to describe the social acceptability of mining operations, the term has since been applied to explore the broad acceptance that communities and other stakeholders provide to the activities of the forest, agriculture, and energy sectors. It is found that human trust (faith), fairness, and

A. K. Soni and P. Nema, *Limestone Mining in India*, Materials Horizons: From Nature to Nanomaterials, https://doi.org/10.1007/978-981-16-3560-1_3

governance may justify the development of more sustainable relationships between industry and society. This section thus presents the concepts practised in industry over the last two decades.

Day by day, mechanized opencast limestone mines are becoming larger, deeper, and more costly to run. Continuous efforts are needed to improve upon the economics of opencast mining activities so that the cost of production per ton of limestone can be brought down and kept contained in a profitable range. Advances in the equipment size, equipment design, and the use of innovative techniques and practices are some of the ways to achieve these economical targets. *Clean technology options* can be added in this list because today our society demands and accepts only those methodologies of mineral and ore excavation that is cleaner, environmentally friendly, and less harmful for workers' health. Considering this, a brief description of clean technology is also covered in this section. The *clean technology approach* includes action to protect the environment by using modern instruments; advanced, remote-controlled, and environmentally friendly machines; and other technological advancements that are feasible for use in the limestone mines of India.

3.1.1 Clean Technology Mining Options

Following proven and established *cleaner technologies* can improve the environmental conditions of Indian limestone mines and mining areas if adopted. Considering their suitability to mining sites, these can be easily put into practice.

- Application of surface miners (continuous mining technology)
- Adoption of wet drilling or dust filtering in drilling at source
- Use of bulldozer ripper as an alternative to drilling and blasting
- Use of hydraulic hammer/rock breaker for secondary blasting
- Application of recent development in blasting technology.

These cleaner technologies for Indian limestone mines are discussed here in detail.

Continuous Mining Technology

Surface miners that eliminate conventional drilling and blasting methods and provide continuous production are dealt separately in detail in Sect. 2.2 of Chap. 2. Surface miners in limestone extraction are being used at selected places in India.

Wet Drilling or Dust Filtering in Drilling

Dust is a major problem in limestone mines. Drilling and crushing sites are the main sources for dust generation. Today, manufacturers are designing their equipment with dust-suppression arrangements. Limestone producing companies should only procure those machines that are attached with such devices despite the higher investment cost. Sometimes locally developed systems (as shown in Fig. 3.1) also prove effective in dust suppression and bring about significant cost savings and environmental risk reduction. In order to reduce the negative and ill effects of dust, wet

Fig. 3.1 A dust-suppression system developed locally at GCW, Kovaya, Gujarat. *Source* UltraTech Cement, 2017

drilling and water sprinkling are other common solutions, but one big problem is that a huge quantity of water is required. Hence, a dust filtering or dust-suppression systems in the machinery should be deployed to maintain a safe and healthy environment.

An environment management cell should be set up and made functional in all limestone mining companies as an in-house facility to monitor the environment quality in and around the mine area and to take necessary steps for protection of both the environment as well as worker's health. Steps so taken will enhance work efficiency for the mine employees and the neighbouring community.

Ripper/Dozer an Alternative to Drilling and Blasting

Ripper/dozer of various makes and brands are available in the market as part of heavy earth-moving equipment. The ripping and dozing tool also comes as an attachment to the excavator. When such tools are attached, the excavator becomes a ripper and

Fig. 3.2 Bulldozer ripper in operation at a quarry face

dozer. This engineering machine is operated with diesel and is mounted on tyres or as a crawler, depending on the ground conditions to be handled, for high mobility and manoeuvrability (Fig. 3.2). Quite often, the term word bulldozer/ ripper (slang) is used for these machines.

Ripping and dozing is an alternate method of excavation that has to be integrated with conventional loading, crushing, and transportation arrangements. This method has limitations because it is not suited to achieve higher levels of production (i.e., >1000 TPD) for many operations. Moreover, while this alternate method reduces hazardous blasting operations, it is inflicted with periodic long-term breakdowns of machines, causing heavy maintenance cost of the equipment and problems and inconvenience in normal mine production.

In general, the choice of mechanical means of excavation (i.e., ripping and dozing as well as using rock breakers) is influenced by several factors and conditions as described here.

1. Topography
2. Mining ground conditions (e.g., rock or ore type encountered and slope condition)
3. Life span for operation of mine
4. Natural restrains on account of site altitude, site climate, etc.
5. Blasting restrictions and waste disposal practices of mine
6. Financial bottlenecks or cost restrictions
7. Other infrastructure available with mining organizations.

Hydraulic Rock Breaker (Rock Breaker or Hydraulic Hammer)

The *hydraulic breaker* is an attachment used for an excavator. Sometimes it is also referred as *rock breaker*. This tool is used for various applications, such as rock or boulder breaking, loose ground excavation in mines, and demolition of construction in civil engineering. The hydraulic breaker, when attached to excavator, becomes a full, heavy-duty piece of engineering equipment. It consists of a machine that is turning the hydraulic energy (supplied by a positive displacement pump) into mechanical energy in terms of percussions against a chisel. The chisel provides alternating impacts to break rock or crumble a certain material by means of the percussive action provided. The rock breaker is a versatile hydraulic machine and indispensable equipment that is economical to use (cost-wise). The working principle of the rock breaker is simple and based on 'percussive impact' principle. Through reciprocating motion of the piston and push chamber, a high hydraulic pressure (power) is generated to give an impact on rock. The striking of the chisel, attached at the front of excavator, causes the breaking of rock and material. These machines alternate means of providing clean production but are not suitable for very large production.

Recent Development in Blasting Technology

Presently, blasting for rock excavation, overburden removal, or limestone production is in vogue in Indian limestone mines and applied as *conventional* unit operations in mining practices. Considerable changes have taken place in mine blasting and mine explosives in recent years. New explosive developments and new blasting techniques mark the recent trend in the area of rock breaking. Therefore, a brief account of these is covered under this section as *clean technology options*. Technological solutions for rock excavations, using blasting techniques for various limestone mines of India are available indigenously and some of the mines are using these latest technologies.

Explosives for watery conditions; non-destructive explosives (expansive chemicals for rock breakage); explosives for hot-hole blasting; new high-explosive initiating devices, such as electric blasting caps (EBC), exploding bridge wire (EBW) detonators, nonel cords, primers, detonating cords, and primadet); plasma blasting technique (PBT); and noiseless propellant-based blasting agents are more recent additions in the area of explosive and blasting techniques for mines. Compared with conventional blasting methods, PBT for rock fracturing is an eco-friendly and clean technique that causes less vibration, less noise, no fly rocks, and dust that uses no chemical substances [1, 2].

The Central Institute of Mining and Fuel Research (CSIR-CIMFR), a national laboratory in mining, has developed tailor-made controlled blasting technology for open-pit applications. Controlled blasting in various limestone formations had fetched substantial and consistent revenue for the limestone and cement industries and tangible benefits in terms of improved productivity to mining companies. These user-friendly blasting techniques ensure total control for all of the environmental problems due to blasting and generate confidence among local communities in the vicinity of mining projects. As a result of *controlled blasting practices*, the workplace safety has gone up significantly.

The controlled blasting methods include:

- Line drilling
- Bottom hole initiation
- Presplitting
- Smooth blasting.

Other more safe, productive, and innovative blasting techniques include (a) the bottom hole decking technique, (b) the effective delay scattering technique, and (c) the bottom hole shock relief technique. These new and recent developments in blasting are site-specific. A short description of each is given here:

(a) The *bottom hole decking technique* consists of air decking at the bottom of the hole in dry holes constructed by means of a wooden spacer or a PVC pipe for improving fragmentation, reducing specific charge, and preventing ground vibrations, overbreak and blast damage in open cut blasting. This reduces peak particle velocity by 40% and overbreak or blast damage reduction by 35%. It is an effective technique for improving blasting productivity as well as safety.

(b) The *effective scattering delay sequence method* involves the delayed firing of blast holes in such a way that the destructive interference of vibrations takes place. This results in reducing the vibration intensity (i.e., peak particle velocity), which is essential in measuring blasting near sensitive structures. This delay sequence scattering method causes a significant reduction in vibrations (20–25% of peak particle velocity).

(c) The *bottom hole shock relief technique* is practised in mines and civil engineering infrastructure projects where excavation is involved. This method consists of inserting reinforced concrete balls at the bottom of blast holes for reorienting the shock energy in the desired directions. This shock energy relief blasting technique greatly reduces the blast vibration intensity by 50% and the overbreak or blast damage by 40%.

Each of these three new controlled blasting techniques are very simple and easy to practice and have proven to be very effective for safety compliance and productivity enhancement. Considerable cost savings can be realized by mining companies engaged in the production of limestone.

Here, the term *blast optimization* is introduced, which means achieving the best blasting operation. Proper optimization of drilling and blasting contribute significantly towards the profitability. Therefore, optimization of these parameters is quite essential. An optimum blast is also associated with the most efficient utilization of blasting energy, less debris from thrown material, and the reduction of blast vibration, resulting in greater degrees of safety and stability to nearby structures [3, 4]. A blast design optimization pyramid is shown in Fig. 3.3 [5].

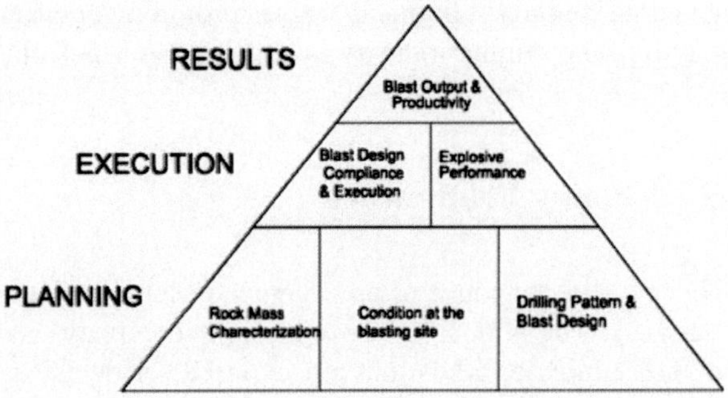

Fig. 3.3 Blast design optimization pyramid

3.1.2 Society and Industry

To produce minerals, the role of society is extremely important. Society and industry have to work in unison and remain responsible and accountable for the environment. Several cement manufacturing companies promote *rural development* for improved quality of life (QOL) in and around their captive limestone mines. Setting up a *cement foundation* for rural development and improvement of the environment of the area—especially for economically backward communities—is a significant step towards the enhancement of the standard of living at a grassroots level. Such foundation projects are based on a national watershed implementation, via the Mahatma Gandhi National Rural Employment Guarantee Act (MANREGA), by the Indira Awas Yojana (IAY), with state health schemes in addition to collaborative projects with the District Rural Development Agency (DRDA). Agriculture related demonstrations for rural areas are arranged at various schools and colleges to promote new varieties of seeds and fertilizers and advances in horticultural farming, animal husbandry, and wasteland development, meeting the requirements of the local public. The rural development activities may also comprise of launching literacy campaigns, checking dam construction, recharging groundwater wells and overseeing their construction, installing drip-irrigation facilities, distributing drinking water in surrounding villages, constructing village roads, providing soil and water conservation programmes, installing biogas plants and smokeless chullah in villages, etc. All of these activities will generate additional employment to local people and improve QOL in mining areas, enhancing the earnings of rural families that are below the poverty lines.

No industry can afford development at the cost of the environment and in particular its impacts on society and socioeconomic fabric. Thus, social dimensions of mining activity cannot be left out. The quality of life must be raised slowly and unabatedly. It is also evident and clear that performance pressures on industrial organizations can be met suitably only when the associated society accepts such operations so there is no industrial unrest. In many ways, the *social licence* reflects the evolving nature of the relationships between industries, stakeholders, and the members of society.

A lot can be done for improving rural development at a grassroots level that includes the principle of sustainable development. In real-life practices, raising the general standard of living as well as providing environment protection should be the motto of the mining industry in general and limestone mining in particular.

3.2 Policy Framework

Nearly, all companies of an organized sector maintain a *corporate policy* for their business. The *policy framework* of any company is a document that contains a set of procedures or goals that are required for decision-making for, production planning and preparation of overall guidelines to run that business in consonance with the national policy framework and the company's profitability. Based on the main policy framework, a more detailed set of policies (e.g., environmental, IT, safety, and security policies) are generally formulated for the efficient and smooth running of the organization. It is quite obvious that a company with a robust, in-place, and sound policy framework can yield better production. A good upkeep of equipment, machinery, and infrastructure; market forecast (assessment of demand and expectations); inventory control; human resources; manpower availability; and insurance of standardized practices within a timeframe are possible with a diligent policy. Not only this, a sound policy framework also covers risk factors involved and maintains an industrial harmony as well.

A mine is a production unit hence, productivity is linked with it. In order to maximize productivity, every mine or mining company needs a meticulous *production plan*. Together with an effective production plan and good scheduling, it is ensured that materials, equipment, and human resources are available when and where they are needed to cover a wide variety of mining activities. Production planning is like a roadmap because it helps one to know where to go and how long it will take to get there. The advantages associated with effective production plan and scheduling are listed here.

(a) Reduced labour costs
(b) Waste reduction
(c) Elimination of unproductive time/ wasted time
(d) Improved process flow
(e) Reduced inventory costs
(f) Optimized equipment usage
(g) Increased capacity of heavy earth-moving machines (HEMM)
(h) Improved on-time deliveries of products and services
(i) Enhanced production quality.

Once it is determined that the criteria to start production are in place, manager will need to communicate the production plan to the employees who will implement it. Key elements of production will ensure an uninterrupted flow of work as it unfolds. Material ordering, equipment procurement, bottlenecks, human resources acquisitions, and training would commence with mining production and lead towards the target.

In brief, effective planning and sound policies coupled with an understanding of key activities by entrepreneurs and business managers that need to be monitored are the vital parameters for the success of an industrial activity. In the following paragraphs, only two aspects of policy framework, namely limestone related industry-specific standards (Sect. 3.2.1) and ISO certification (Sect. 3.2.2) for the captive mine of limestone, are highlighted.

3.2.1 Limestone-Related Industry-Specific Standards

Environmental standards form the most apparent source of pressure for organizational action in environmental protection. In general, we consider these standards only in their regulative perspective, and they remain symbolic. However, many times they become strong enough for contest in the court of law. The needs for industry-specific standards for limestone mining have been realized long since because of the site-specific and dynamic nature of the mining industry. Of late 'Industry Specific Standards for Limestone Mining in India' was developed by CSIR-CIMFR through a Central Pollution Control Board (CPCB) sponsored project [6]. These standards were formed based on a study of 15 limestone mines across the length and breadth of the country. The *newness* of these proposed environmental standards (Table 3.1) are as follows.

- For air environment in place of SPM or RPM, PM_{10} & $PM_{2.5}$ are included and their limit prescribed or assigned based on interaction with various agencies involved (e.g., CPCB, industry, and concerned ministry).
- To deal with air pollution including fugitive dust concentration, environmental standards are developed based on activity and workplace. These are effective within the mines and also at vulnerable parts of mines where workers are exposed to dust for a maximum period of their work hours.
- For water environments, selected parameters are taken into consideration for quality evaluation. Also, effluent discharge into streams and rivers is monitored periodically and necessary remedial measures taken.
- For noise, DGMS norms are adopted in order to avoid conflicts with existing environmental laws. The standards have been created for industrial surroundings during day and night time.

The framed standards (proposed) can be applied at various levels of their route between the source of contamination and its target to attain the desired environmental quality. In brief, these standards are meant to protect human health from adverse effects and to ensure well-being of the common workforce engaged in limestone mining activities in particular.

Table 3.1 Environmental standards for limestone mining in India (proposed)

(a) Proposed fugitive standards for air quality at workplaces in limestone mines[a]

Activity	Maximum limit: 30–35 m distance downward direction from the activity for eight hours of monitoring during mining	
	PM_{10}(μg/m^3)	$PM_{2.5}$(μg/m^3)
Drilling	800	480
Loading and unloading	800	480
Hauling*	800	480
Waste dumping	800	480
Crushing	800	480

(b) Wastewater discharge standards for limestone mines (proposed)[b]

S. No	Parameter	Value
1	pH	6–9
2	Total suspended solids (TSS) (mg/l)	100
3	Oil and grease (mg/l)	10
4	Total dissolved solids (TDS) (mg/l)	2100

(continued)

Table 3.1 (continued)

(a) Proposed fugitive standards for air quality at workplaces in limestone mines[a]

(c) Noise-level standards for limestone mines (proposed)[c]

S. No	Parameter	Noise limits	
		Day time (6:00 AM to 10:00 PM)	**Night time (10:00 PM to 6:00 AM)**
1	Noise level (Leq)	75 dB(A)	70 dB(A)

[a] *Note* (1) Any other activity refers to mining or related ancillary activity that is directed towards limestone production. (2) The air quality monitoring shall be done for an 8 h duration minimum with prescribed standard instruments as referred in National Ambient Air Quality Standards (NAAQS)

*Emission monitoring should be done adjacent to haul roads

[b] *Note* Other water quality parameters are not significant in limestone mining

[c] *Note* Noise levels shall be monitored both during day and night times on the same day while in operation

3.2.2 *ISO Certification*

The International Organization for Standardization (ISO) is an independent, non-governmental international organization with a membership of 164 national standards bodies across the globe. Since 1946, the ISO has published over 22,636 international standards covering almost all aspects of technology and manufacturing (https://www.iso.org/). *ISO international standards* ensure that products and services are safe, reliable, and of good quality. For business purposes, they are strategic tools that reduce costs by minimizing waste and errors and increasing productivity. The ISO standards outline priorities for the coming years through continual improvement of the existing system. They help companies to access new markets, level the playing field for developing countries, and facilitate free and fair global trade. By providing guidance and strategic direction, they allow businesses to prepare for a future where constant change will require upgrades. Such certification is a major help.

ISO certification for a captive opencast limestone mine of a cement plant requires voluntary efforts by the company that owns both the mine and the plant. In doing so through its members, it brings together various experts to share knowledge and develop a consensus-based and market-relevant product of good quality. The results will support innovation as well as international norms that provide solutions to global challenges.

Some popular and well-known international standards relevant to opencast limestone mines and mining are

- ISO 9000 series—Quality Management
- ISO 14000 series—Environment Management Systems
- ISO 26000 series—Social Responsibility
- ISO 27001 series—Information Security Management
- ISO 31000 series—Risk Management
- ISO 45000 series—Occupational Health and Safety Management Systems.

The very purpose of describing ISO certification and standardization in the policy framework section of this book is that the profit-oriented private sector is either not aware or not willing to spend for quality. We feel that the cost associated with ISO certification in limestone producing companies of developing countries, which lag in environmental management, is justified through this approach. If the care for quality is not maintained, the scope of company growth gets narrowed. Therefore, limestone mines—concerning either cement plants or industry—should and must have a strong policy for maintaining quality standards throughout their production life. If one becomes aware and implements ISO standards into practices, the objectives of improved environment, health, and safety are easily fulfilled.

Since the ISO addresses the reduction of waste production and improvement in environmental footprints, such companies have an ability to adopt 'greener' practices. Quality management (QM) and environmental management (EM) are business practices that support and benefit companies in their daily operations. Similarly, environmental risk and health are also integrated with it. Thus, sound policy development applied towards certification is a laudable step of the operative mine, plant, or company.

In brief, it is apparent that ISO standards are voluntary, and acquiring this certification has a marketing advantage that is necessary for fruitful competitiveness. However, it is observed in developing countries that it is difficult to pay the high costs associated with such certification, which causes significant disadvantage to the companies in the global marketplace [7].

3.2.3 Economic Policies and Environment

In the national and regional context, the macroeconomic and sector policies govern the use of mineral resources and resulting products, which includes limestone. The complex interrelationship between the economics, policy, and environment is not well understood by everyone. Hence, the ideal approach to forming such policies has been that of general equilibrium. In developing countries, such equilibrium is selectively feasible wherever data and skills are available, may be due to societal pressure or environmental awareness. To a limited extent, the environmental pollution and exploitation of the resources in developed nations have been brought under control. Market pressure and policy distortions leading to industrial unrest are also kept under check by economical and industrial policies. This has necessitated periodical review of the regulatory framework, providing reforms to promote efficiency and productivity in mines, reducing poverty in mining areas, and implementing improved environment practices.

Economical and environmental policies are linked with global atmospheric changes too [8]. Today, we observe many forms of capitalization, global marketization, and climatic challenges. In the mining and mineral industries, integration of economic policies, management, and environmental issues have been recognized in the late 1990s in the form of *environment management capacity building* (EMCB). This technical assistance is provided by the government of India and has addressed both population-based economic order and need-based sustainable societal development of mining areas [9].

Limestone quarries and their restoration provide a major opportunity to protect and enhance biodiversity. Most of the limestone captive mines of the cement industry have a long legacy of good-quality restoration. Therefore, one can make a significant contribution in terms of environment preservation and protection. In this regard, some key areas are given here.

- Flora and fauna protection and improvement in the affected mining land of rural areas (i.e., *wildlife protection* and *biodiversity enhancement*)
- Designing and creating new landforms and habitats to support local biodiversity
- Soil improvement (i.e., converting low-nutrient habitats to high-nutrient habitats).

Proper planning and implementation of these measures will succeed in the restoration of the environment in limestone areas. In particular, only green mining [10] should be encouraged.

3.2.4 Some Other Aspects of Policy Framework

Impacts from the highest limestone consuming industry (i.e., cement industry on greenhouse gases emission (CO_2), carbon footprint generation, and climate change) are quite high. Hence, a sound policy framework is required within the organization. It may be noted that the cement manufacturing process is resource and energy intensive due to the extreme heat required. Depending on the type and process, a cement plant requires 60–130 kg of fuel oil and 110 kWh of electricity to produce one tonne of cement. For every tonne of cement made, the process releases approximately 01–1.25 tonnes of carbon dioxide. The reason for such a high value of CO_2 generation is that 'limestone burning' (a rock to be heated up to the 1450 °C) and 'kiln firing' both are an extensive energy-consuming processes. Thus, more the energy consumption higher will be the carbon dioxide generation.

Captive limestone mining operations are divert from lowering the carbon footprint and promoting climate change by considering that it will not matter significantly. The fact is that if produced limestone contains higher percentage of moisture—whether inherent or during handling—it will require more energy to make *dry clinkers*. (When large amounts of fossil fuels are used to heat a high temperature kiln to around 1400 °C, limestone and other raw materials are decomposed to form a substance called clinker, which is then combined with gypsum to make cement.) Nearly 10% of emissions come from fuels needed for limestone mining and transportation of the raw materials.

Thus, it is clear that emissions from cement manufacturing depend largely on the proportion of clinkers used in each tonne of cement [11]. The type of fuel and efficiency of equipment used during clinker production also have an impact. Development of *clinker free cements* could reduce emissions significantly. Those firms developing clinker free cements claim that nearly a 90% emission reduction is possible. Therefore, policy intervention at the mining level becomes helpful and needed by the industries.

Around half of the emissions from cement making are *process emissions* (40%), arising from the chemical reaction Eq. (3.1), which cannot be eliminated by changing fuel or increasing efficiency).*

$$CaCO_3 \rightarrow Cao + Co_2 \tag{3.1}$$

**Nearly 40% of emissions come from burning fossil fuels to heat kilns to high temperatures needed for this calcination process. This is one major reason why CO_2 emissions are often considered difficult to cut in cement industry.*

The cement industry consumes the highest quantities of limestone accounts for 5–6% of global greenhouse emissions, and its actions are crucial to helping achieve the central goal of the 2015 Paris Climate Change Agreement to limit global warming well below 2 °C and as close as possible to 1.5 °C. The World Cement Association (WCA) is urging industry members to increase efforts to adopt new technologies quickly and at scale to reduce its CO_2 emissions in order to effectively help fight climate change (Fig. 3.4). The WCA has called upon the entire cement industry to put greater focus on innovation in order to make crucial progress on reducing CO_2 emission (https://unfccc.int/news/world-cement-association-urges-climate-action).

Technologies currently deployed by the cement sector only deliver 50% of the CO_2 savings required to achieve the Paris Agreement goal which is typically very slow. Today companies are increasingly serious about reducing their CO_2 emissions to help policy-makers in climate change support. An inspiring example of climate action in the cement industry can be found from India's Dalmia Bharat Cement, Ltd. This company is a leading cement manufacturer undertaking measures to use eco-friendly raw materials and to replace natural resources by using the waste generated from other industries. With a group carbon footprint of 493 kg CO_2 /tonne of cementitious material, this company delivers one of the least carbon-intensive cement operations in the world. From 2015 to 2016, the Dalmia Cement company said it achieved a 16% reduction in its carbon footprint. Last year, Dalmia Cement also became the first cement company to join RE100, which is a global initiative of the world's leading companies committed to 100% renewable power (https://unfccc.int/news/world-cement-association-urges-climate-action).

From the life cycle assessments study of the mining and mineral processing of iron ore, bauxite, and copper concentrate carried out by Norgate and Haqueit [12], it was observed that loading and hauling made the largest contributions (approximately 50%) to the total greenhouse gas emissions from the mining and processing of iron

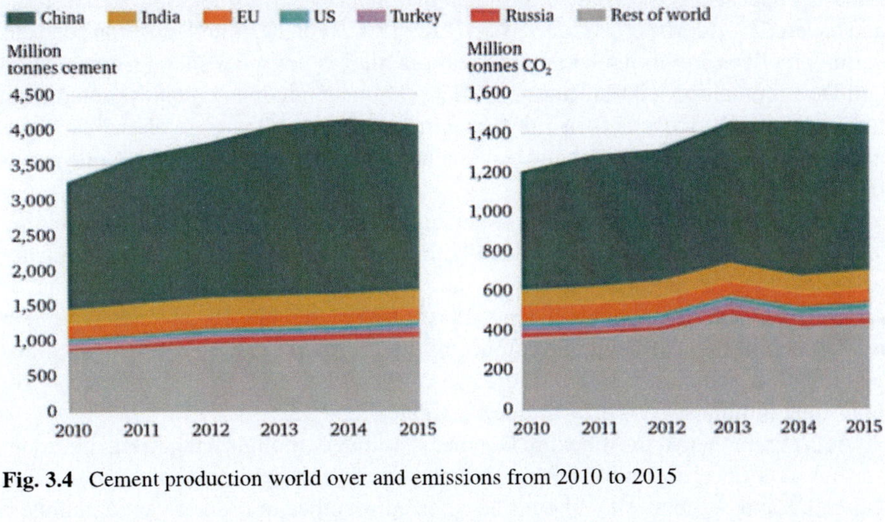

Fig. 3.4 Cement production world over and emissions from 2010 to 2015

ore and bauxite (11.9 and 4.9 kg CO_2 e/t, respectively). In the case of copper ore, it is the crushing and grinding steps that make the largest contribution (approximately 46%)—particularly the grinding operation—to the total greenhouse gas emissions for the mining and processing of copper ore (628 kg CO_2 e/t concentrate). This analysis was mainly completed for Australian metallic ore mines and points towards a fact that the limestone mining industry may have similar culprits of greenhouse gases and CO_2 emissions. However, a detailed study for the entire limestone sector (not the cement sector alone) is still required to help form future scopes of research.

3.2.5 Limestone Mining: Statutory and Legal Compliances

As explained earlier in Chap. 1, Sect. 1.2.1, in India limestone, is classified as both *minor* and *major* minerals, according to its use. For commercial and industrial purposes, limestone mining is largely covered under the major mineral category. For all legal compliance of limestone mining, the Ministry of Mines (MoM) of the Government of India, the Indian Bureau of Mines (IBM), and the Geological Survey of India (GSI)* are the concerned and responsible organizations of India. Here, *legal compliance* and *statutory compliance* are synonymous for the description purpose. This includes legislation, administration, policy formulation, etc. for the entire limestone mining industry. Wherever limestone is a minor mineral, the power to frame policy and legislation and statutory compliance are entirely delegated to the respective state governments (through state directorates or a department of geology and mining) where limestone mining is done. For limestone as a mineral, the mineral legislations of the country are applicable. These must conform to the provisions of the Mines and Mineral Development and Regulation (MMDR) Amendment Act of 2015 (original act, MMDR Act of 1957). For all of the limestone mines of India, mineral concessions are granted to Indian nationals or entities incorporated in India and covered by the Minerals Concession Rules (MCR) of 2016, and limestone conservation practices are implemented into real practice as per the Mineral Conservation and Development (MCDR) Rules of 2017. Various amendments incorporated from time to time into these acts or rules remain applicable at the time of enforcement, and the IBM is the leading agency responsible for their execution in limestone mines throughout the country.

The legal framework for the limestone mining sector—from exploration to mining and until closure—is in place and well defined. An authentic record of the Ministry of Mines (MOM) delineating salient details of limestone mining sectors can be referred to in addition to other relevant legislative documents [18]. If an entrepreneur or investor desires to begin limestone mining in India, some requirements must be complied, which are referred to as *legal compliances.* These requirements for limestone mining sectors are provided in Table 3.2.

At the end of book, in Annexure D, other statutory compliance requirements under various acts and rules have been listed for mines and environmental areas. Both Table 3.2 and Annexure D should be referred together in order to get a complete

Table 3.2 Legal compliance and concerned authorized agencies for limestone sectors

S. No	Item	Concerned authorized agency for legal compliance
1	Establishment of company or enterprise	• Min. of industry, GOI • Min. of mines (MOM), GOI • Min. of corporate affairs, GOI
2	Permission for prospecting (i.e., prospecting licences	Min. of mines and IBM
3	Geological assessment of limestone reserves through detailed exploration	• GSI and MECL • Min. of mines
4	Statutory clearances before start of mining operation (premining) • Diversion of forest land for mining of limestone under Forest (Conservation) Act, 1980 and forest clearance • Wildlife clearances (sanctuary, reserves or special zone permit and clearances) • Environmental clearance (for more than 05 ha of mining lease area) • Gram Sabha consent • Consent of surface rights from revenue records	• Min. of mines (MOM) • Min. of environment, forest, and climate change (MoEFCC) • State government and local city or village corporation (panchayat) (Also refer to Annexure D)
5	Obtaining mining lease for mining after land acquisition by the company, organization, or entrepreneur	• Min. of mines and IBM • State government
6	Consent to operate mine (including permission for safety zones and safeguards of person or labours and mining near protected areas)	• Min. of mines and IBM • State government (for minor minerals) • DGMS, Min. of labour
7	Mine plan and mining scheme preparation and approval	• Min. of mines and IBM • State government
8	Groundwater clearance	• Min. of water resources • CGWB/CGWA (central ground water board/central ground water authority) • IBM and state government
9	All types of statutory clearances along with returns to be filed periodically to operate limestone mines • Various provisions under Environment Protection Act, 1986 (amended in 1991) and Environment Protection Rules, 1986 • The Air (prevention and control of pollution) Act, 1981 • The Water (prevention and control of pollution) Act, 1974	• Central pollution control board (CPCB) • State pollution control board (SPCB) (Also refer to Annexure D)

(continued)

Table 3.2 (continued)

S. No	Item	Concerned authorized agency for legal compliance
10	All types of statutory returns (monthly and annual) under MCDR and MCR	IBM (through MTS)
11	Explosive licences and all types of statutory provision and consent to use and store explosive under Explosive Act, 1884, and Explosive Rules 2008 (amended in 2017)	Petroleum and Explosive Safety Organization (PESO), Min. of commerce and industries, government of India
12	Approval of closure plan	IBM (through MTS)
13	All types of statutory provision for royalty and dead rent, etc.	• State government • District and state authorities concerned
14	All types of statutory provision for taxation (i.e., GST or income tax) as per the rules applicable for industry	Central or state government
15	Approval for railway siding (if limestone is transported by rail)	Min. of railways or DRM of concerned railway zone
16	Approval for storage of diesel in bulk	Min of petroleum/PESO
17	Official registration or enrolment with *District Mineral Foundations* (DMF) #	State government or district and local authorities concerned
18	Official registration with *National Mineral Exploration Trust* (NMET)@	National, state, district, and local authorities concerned

Note Min. refers to Ministry

#**District Mineral Foundations (DMF)** are statutory bodies in India established by the state governments by notification. They derive their legal status from Section 9B of the Mines and Minerals (Development and Regulation) Act of 1957 as amended on 26 March, 2015, in the Mines and Minerals (Development and Regulation) Amendment Act of 2015. Each District Mineral Foundation (DMF) is established by the state governments by notification as a trust or non-profit body in the mining operation-affected districts

@**The National Mineral Exploration Trust (NMET)** was established by the Government of India in August 2015, in pursuance of subsection (1) of Section 9C of the Mines and Minerals (Development and Regulation) Act, 1957, with the objective to expedite mineral exploration in the country

NMET operates its office from the Ministry of Mines, New Delhi. NMET supports regional and detailed mineral exploration in the country and other activities approved by its governing body. They include special studies and projects to identify, explore, extract, improve, and refine deep-seated and concealed mineral deposits, studies on mineral development, sustainable mining, mineral extraction, and metallurgy by adopting advanced scientific and technological practices, detailed and regional exploration for strategic and critical minerals, upgraded mineral exploration status in an area from G3 to G2/G1 levels, exploration leading to grants of mineral concessions, aerial and ground geophysical surveys, geochemical surveys, and increasing the number of personnel engaged in mineral exploration

To implement mandated activities, a National Mineral Exploration Trust (NMET) fund has been established. The NMET fund receives money from holders of mining leases or prospecting license-cum-mining leases in an amount equivalent to two percent of the royalty paid via terms of the Second Schedule of the MMDR Act idea of different types of procedural requirements for a limestone mine to operate. It should be mentioned here that with time and amendments, these statutory and legal compliances keep changing.

3.2.6 *Mining Tenement System (MTS)*

The very purpose of writing this section is to make the reader aware of the existing legal compliance system that has been primarily conceptualized to smooth the link between the mining industry and statutory bodies in India, such as the Indian Bureau of Mines (IBM), which is an important core agency for the limestone mining sector. An inquisitive mind can learn more about MTS and how it functions.

The Mining Tenement System (MTS) is a digital repository of entire *life cycle analysis* of each mining concession granted for limestone as major minerals but is not included when treated as minor minerals. Thus, MTS is an *online information system* on limestone mineral resources of the country that have a graphical information database (GIS) as well as information in textual form. MTS includes automation of all the processes and approvals at various levels of governance, thereby mapping all of the activities from identification of potential mineral-bearing areas to post-closure of mining activity. This enables real-time transfer of electronic data and files with the geographical information system (GIS) interface.

The IBM has been the implementing agency (IA) for MTS since 2016. Its key activities are depicted in Fig. 3.5. In brief, the MTS system has been primarily conceptualized to transform the functioning of the IBM for efficient, effective, and transparent delivery with a provision to engage the state mining departments all across the country as per the option exercised by them. This online computerized system provides information in visual graphic form in GIS and in a textual form known as a *registry component*. The information database to be maintained in the

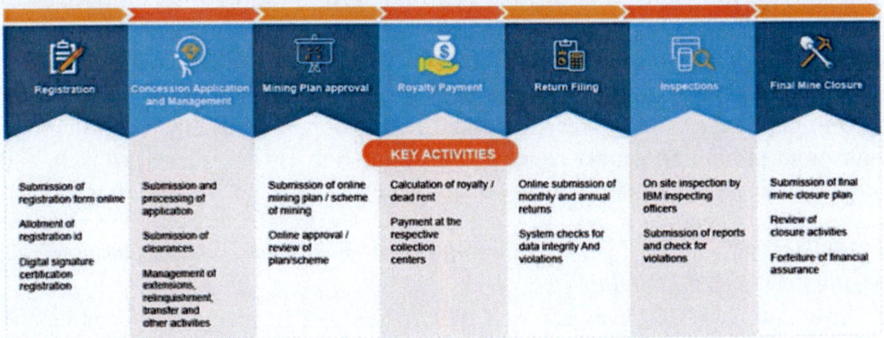

Fig. 3.5 Key activities of MTS. *Source* https://mitra.ibm.gov.in/Pages/IBM_Home.aspx

project includes the duration of concessions, security of tenure, criteria for grant of mineral concession, transferability of PL, reservation of areas, details of grant of mineral concessions, disputes, forest details, MOEF clearance, infrastructure, taxation, illegal mining, captive mines, export–import, etc.

The computerized online register of the Mining Tenement System (MTS) is aimed at providing ease in the implementation of the National Mineral Policy. The legal framework at central and state levels, various procedures for granting mineral concessions, stakeholder analysis, workflow analysis, functional requirements, specifications, solution architecture, change management and capacity building strategies, implementation strategies, and financial estimates can be handled with MTS online.

3.3 Corporate Social Responsibility and Its Role

Corporate social responsibility (CSR) is a business approach that contributes to sustainable development by delivering economic, social, and environmental benefits to all stakeholders. The purpose of CSR is to drive change towards sustainability. The World Business Council for Sustainable Development (WBCSD) is a leading international umbrella organization for sustainability promotion and CSR-related matters that developed a clear understanding of corporate social responsibility. There is an increasing focus by the government, public sector organizations, and private sector cement companies to examine their social responsibilities and to know how their industrial activity impacts society.

In India, the CSR clause comes under Companies Act of 2013, and the Ministry of Corporate Affairs (MCA) of the government of India is the custodial agency. For Indian companies, the CSR Rules of 2014 have been notified by the MCA, and Section 135 and Schedule VII of the Companies Act of 2013 is promulgated.

Now let us understand, what is meant by corporate social responsibility (CSR). Responsibility for what and to whom, and who is calling for companies to be socially responsible? We will be focusing our description for the cement sector businesses and the community (i.e., cement manufacturing), which are value-added products and materials from limestone, in the next section. The CSR activities that can be undertaken by a company to fulfil its obligations are poverty alleviations (i.e., providing food to the hungry and to prevent malnutrition); supporting healthcare; promoting education; promoting gender equality; setting up homes for orphans, women, and senior citizens; supporting animal welfare; safeguarding the welfare of scheduled castes (SC), scheduled tribes (ST), and other backward classes (OBC); protection of India's national heritage and promoting rural development—all guided towards *environmental sustainability.*

In literature, CSR topics are well described [13–15], giving the ideas behind CSR, its definition(s), theories (social contract's theory and legitimacy theory), stakeholders, indicators, and ways of assessing and reporting its performance to the industry in general. Our purpose here is not to describe all these different aspects of CSR but to make our point that CSR is an inescapable priority for business leaders in every country, including India [15]. The CSR-associated concepts are for the benefit of society as a whole, including the environment, companies, activists, stakeholders, local villagers, and entire communities. The only thing that is needed is implementation and concrete actions of CSR that should be considered by the industrial organization or business houses.

3.3.1 CSR and the Indian Cement Sector

In India, the World Business Council for Sustainable Development (WBCSD) and the CII-ITC Centre of Excellence for Sustainable Development (CII-CESD) provide fact-based information on the CSR legislation in a detailed manner. India's corporate law received a major upgrade with the enforcement of Companies Act of 2013, which came into effect in April of 2014. The intent of this act is to improve accountability and responsibility of companies regarding business conduct. The Companies Act makes CSR a matter of corporate governance from planning, monitoring, and reporting perspectives. The act mandates 2% spending on CSR activities. The WBCSD and the CII-CESD will monitor developments and suggest improvement to obtain the best practices.

> *Indian corporate sector companies came forward to make use a part of their profits to support social and economic changes and give back to the people of region they are working with. In cement sector, the UltraTech Group (Aditya Birla Company) has proactively championed social responsibility in cement sector. A significant percentage of the profit share of the company is spent to support healthcare, education, and environmental sustainability activities. As part of their visionary efforts, beyond 2020, company is committed to create a more employable workforce, building a Greener India and innovating for good and green products. UltraTech have trained youths in skills that will enhance their earning potential. Through innovation, new product development always remains in pipeline. In line with greener India commitment, UltraTech Cement is making environmental sustainability a key part of the cement manufacturing process and value chain across their cement plants all over India. In this way corporate, industry, NGO and PAP's (project affected personnel/people) address sustainability challenges in cement sector for which limestone is the main raw material constituent.*

3.4 Limestone Quarrying on a Small-Scale in Society

Limestone mining on a small scale, when discussed as per international norms, brings two words to mind: *artisanal mining* (ASM) and *small-scale mining* (SSM). Both of these are difficult to differentiate, so they depend on individual perception [16]. For example, when mining operations in an area are carried out on a small scale by an entrepreneur with limited resources, the infrastructure can be categorized as *quarrying*. Here quarrying is a small-scale mining category. Normal mining methods and conventional practices are the main approaches used to excavate limestone. Such artisanal small-scale mines are spread in many states of India wherever limestone is found. Limestone mining in the Sirmour district of Himachal Pradesh and dolomite mining in the Mandla district of Madhya Pradesh are some examples of this category (see https://www.downtoearth.org.in/news/ngt-steps-in-to-save-kanha-tiger-from-smallscale-mines-43945).

SSM in India is generally carried out on acquired mining leases through license(s). This practice is highly labour intensive with no or very little adequate equipment or tools for the excavation of rocks (mineral or overburden). When not done properly, SSM often turned into an illegal mining operation. Since small-scale mining (SSM) is most often accomplished with the involvement of local people, such mines are controversial and politicized. Small-scale mining issues are complex because they are multiangled between environmentalists, miners, village communities, and government. These issues are mostly attached to income, livelihood, and employment issues that are not centred but have severe splits among the various constituents of society. Relationships between environment and pollution or industries and mining become a political football match. The fear of banning mining on one hand and being stalled by litigation on the other hand haunt the stakeholders of SSM. The state exchequer remains perturbed about the revenue loss from small-scale mining. Although some argue positively and some negatively, SSM business continues as usual.

In such small mines, the mining lobbies and mine owner(s) predictably become critical and strong with time. However, the local administration continues to find it difficult to enforce the rules, provoking scepticism over the effectiveness of enforcement agencies. Several state and central agencies are involved in the monitoring and enforcing of laws and regulations.

Unskilled miners pay little attention to mining guidelines, opting instead for methods that appear easy and cost little. Ignoring the norm of systematic open-pit mining or strip mining, they pick and carve the best minerals and leave the rest as waste. Contrary to the rules, this mining is done through a tailor-made approach.

For residents in the limestone belt, mining offers jobs that include loading and transport operations. Further, local people consider (contend) this as vital because dependence on the government is reduced and their land becomes more valuable. Thus, small-scale mining deeply penetrates society at a much faster rate for the obvious reason that the local market is growing stronger. Another reason for its growth is the higher level of literacy by doing mining jobs. The passionate arguments relating to environment-versus-mining remain alive when the exploitation of minerals continues. At the same time, the sword of discontinuance and mining bans also continue. For mining and environment benefits to continue together, more constructive dialogues between mining professionals and environmentalists must be approached. If SSM is stopped, its direct consequences are visibly observed in the society. For SSM, it is unrealistic to say no to mining. Instead, the focus should be on the scientific management of mines. State government officials should stress the importance of scientific mining and the proper rehabilitation of mining sites. An eco-friendly development—if promoted and maintained in the proper manner—can compensate for any revenue losses from mining and cement plants. Thus, for both small-scale quarry excavations and the mining society, a win-win situation is always desirable as well as popular in the long term.

3.5 Contemporary Minerals of Limestone

Dolomite and *limeshell* are the contemporary minerals of limestone. Small mineralogical and chemical changes in limestone, when brought out naturally, turns the calcareous rock (i.e., limestone) from limestone to dolomite ($CaCO_3 \cdot MgCO_3$). The difference in its composition is described in Sect. 1.3. As per the government of India Notification S.O. 423(E), dated 10 February 2015, dolomite has been declared to be a minor mineral; hence, the producers report the production data directly to the respective states and not to the IBM. The Indian Bureau of Mines (IBM) of Nagpur only collects the statistics of its production and lease details, which are reported for the national mineral inventory.

Dolomite theoretically contains $CaCO_3$, $MgCO_3$, CaO, and MgO. However, in nature, dolomite is not available in this exact proportion; therefore, in commercial parlance, the rock containing 40–45% $MgCO_3$ is usually called dolomite. Dolomite rock that contains either calcite or a mixture of calcite and magnesite in addition to dolomite is called *dolomitic limestone* and is considered a contemporary mineral. Dolomite is mostly used for refractory purposes in furnace linings. It is grouped under flux and construction minerals and is important for iron, steel, and ferroalloy industries. Dolomite occurrences are widespread in almost all parts of India.

Similarly, *limeshell* is a calcareous material found in lake and seabeds that is also a contemporary of limestone. These limeshells are found in the lake bed by the accumulation of the dead organism shells. Limeshell contains very low iron content; therefore, it is useful for manufacturing white cement. In India, besides surface mining, lake and seabed mining for such calcareous material is also practised at Vembanad Lake of Kerala [17]. Limeshell found on the beds of water bodies are mined by dredging thicknesses ranging from a few centimetres to several metres. In the sea coast of Gujarat, Odisha, and Maharashtra, limeshell is excavated clandestinely (not reported officially) for economic gains. As per the IBM statistics [18] as on March of 2017, twenty six leases of limeshell were operational from the 2117 leases of limestone available (see Table 3.3) [18]. Surveys of other minerals under the calcareous category (i.e., limekankar, chalk, and clay) show that they have substantial

Table 3.3 Materials contemporary of limestone

Name of mineral	No. of leases	Lease area (ha)
Limeshell	26	3061.99
Dolomite or dolomitic limestone	Production (mining and handling) 7% of limestone	
Limestone	2117	160,012.84

carbonate percentage in them, but they are not contemporary to limestone because of differing properties.

Compared to limestone, there is far less mining and handling of dolomite (dolomitic limestone) (around 7%). As reported in literature, dolomite mining was first commenced in Orissa in 1898 around the Birsa and Rourkela areas. Thereafter, the mining was spread to nearly 10 states of India where it was available in various geological deposit forms. Madhya Pradesh and Odisha are the leading dolomite producing states in India. All of these states produce dolomite from small and large mines with different production capacities.

As clarified earlier, limestone and dolomite are contemporary minerals of one category. Their mining processes are identical with same methods used for excavation (only opencast mining), overburden removal, mechanization, loading, and transportation.

3.6 Production Cost of Limestone Mining Operations

It is difficult to ascertain a single figure for the cost of limestone mining operations because the operational cost varies over a wide range. Cost varies from mine to mine, with the extent of mechanization, and material procured and stocked as per the prevailing local conditions at that mine site. In a limestone mine, various unit operations, such as drilling and blasting, crushing and grinding, transportation of ROM, overburden, material, and mineral handling, are required on a daily basis, which adds to the total cost of the limestone mining operation. This is usually called the *production cost of limestone* for a single weight unit (i.e., Rs/ton). The capital cost incurred by the company and several other costs for exigencies, depreciation, etc. are not included in the production cost. A typical cost breakup of the limestone mining operation in a large mine with both mechanized and manual operations is given here.

Cost of salary or wages = 70%
Cost of material, transport, and various unit operations of mines = 14%
Cost of royalty, cess, DMF, etc. = 03%
Cost of labour and employee welfare = 5%
Overhead and other miscellaneous costs = 8%

Total cost of limestone production/mining operation = 100%
(Excluding fixed asset cost/capital cost)

While mining limestone in an open-pit mine, the costs for drilling, explosive consumption, blasting, mine maintenance, and machinery utilization (including machine performance and breakdown cost) are the constituents of operational costs at the pit head. While calculating the costs at the consumer end, the transportation and other costs as indicated were added [19]. In any limestone mine—whether for captive use of a cement plant or for other uses—an adequate size of the mineral (i.e., limestone or dolomite) with fewer fines is the first necessity of the production. This kind of production certainly involve costs that can be effectively controlled through scientific approaches and research studies. Thus, it is evident that the cost of limestone mining operation is variable from mine to mine and from organization to organizations, but it can be kept in check by the local mine management.

3.6.1 Marginal Cost of Production

Limestone mining is a very competitive (throat-cutting) industrial business. To achieve the best production with positive economic growth—even if a small amount of cost is saved—considerable savings can be made by an individual organization. In this respect, the marginal cost has a good scope.

Let us understand the *marginal cost* (MC) of production in limestone mining. The *marginal cost of production* is the change in the total cost that comes from making one additional item or producing one ton extra (i.e., increased output). The marginal cost of production includes all of the costs that vary with the level of production.

The marginal cost of production is most often used by manufacturers as a means of isolating an optimum production level. Manufacturers often examine and calculate the cost of adding one more unit to their production schedules. This is because at some point the benefit of producing one additional unit and generating revenue from that item will bring the overall cost of production down. The key to optimizing manufacturing costs is to find that point or level as quickly as possible. At this point, manufacturing and production become *profitable.* The purpose of analysing the marginal cost in limestone mining is that the mineral is a low value, so the cost cushioning is very small. This is an important factor in economic theory because a company that is looking to maximize its profits will produce up to the point where marginal cost (MC) equals marginal revenue (MR).

For example, consider a limestone producer with a captive mine and Rs. 75 is the cost of production for each tonne of rock. This cost includes all overheads, industrial expenses, operational cost, cess, royalty fees or taxes, etc., excluding the capital (fixed) cost. If the mine incurs Rs. 100 as fixed costs per month, and if it produces 50 tonne of limestone per month, the fixed cost is 100/50 = Rs. 2 per ton. Therefore, the total cost per tonne of limestone produced (including all expenses) will be

$$\text{Production/tonne} + \text{Capital cost} = \text{Rs.}\,75 + \text{Rs.}\,2 = \text{Rs.}\,77$$

Now, if the mine had increased production volume and raised production to 100 tonne/month, the capital cost will be 100/100 = Rs. 1 per ton. This is because fixed costs are spread out across units of output. In this way, the total cost of limestone production per ton would drop to Rs. 76 (Rs. 75 + Rs. 01); thus, increasing the production volume results in marginal costs to go down. Therefore,

$$\text{Production cost} = \text{Fixed cost(s)} + \text{Variable cost(s)}$$

Therefore, it is evident that marginal cost of production can contribute to more economical production cost of limestone. This is yet another important reason why a company's emphasize is on increasing the production figures of mines.

3.7 Sustainable Alternatives

Currently, there is an increasing trend of promoting underground mining in India. Therefore, the underground mining of limestone is the best sustainable method of mining limestone in the future. Recall that limestone is a mineral of low value found near the Earth's surface. Its current production in India is through surface mines using either opencast (stripping and hill mining) or open-pit mining methods. The need today is not just mining, but mining to improve the environment, maintain safety, and keep the production costs effective in a sustainable manner. Therefore, transforming opencast limestone mining into both a safe workplace and a healthy environment to one of underground mining could be an alternative for limestone mining in the future.

Major underground mining 'players' are looking forward for expanding operations and increasing production in the coming years. Underground mining operations could be a better option for limestone industry, provided cost–economics are checked before making a decision. It is evident that underground mining is less productive compared to opencast. Underground mining is a comparatively tougher method with various risk factors involved. However, it is less damaging to the surface and general environment. The risk factors of underground mining of limestone can be reduced to a minimum—if not to zero—at once when a decision has been made to go underground. Site-specific underground mining methods, depending on the size, shape, and orientation of the ore body; the grade of minerals; the strength of the rock materials; and the depths involved have to be selected for limestone excavation. In this way, the complexities and challenges of underground mining are eliminated, making the transformation of limestone excavation and mining from surface mining to underground mining possible in India.

According to industry sources, underground operation in India is on the verge of transition from the conventional to an automated operation. In the near future, this will allow there to be hardly any deposits left near surface. Hence, the underground operation shall become more prominent.

The future of underground limestone mining also depends on how well we are prepared to address deeper limestone mining challenges. It is true that in spite of the high risk and investment involved, underground mining can become safe and productive with the help and proper use of modern technologies and equipment. With the use of digital technologies, automation, and technologically advanced equipment, underground mining operations will be safe as well as more productive. Underground mining also will provide a good opportunity to explore and exploit the huge deposits underlying in deeper earth layers.

References

1. Huisheng Z, Xinghua X, Yuqing F (2018) Rock breaking methods to replace blasting. IOP Conf Ser Mater Sci Eng 322:022014. SAMSE IOP Publishing. https://doi.org/10.1088/1757-899X/322/2/022014
2. Nantel J (2012) Improvements in mining technology. In: American society of civil engineers (ASCE), fifth international conference on space, Published online, Apr 26, 2012, pp 806–812. https://doi.org/10.1061/40177(207)109. Accessed on 24 Jan 2020
3. Nanda NK (2003) Optimization of mine production system through operation research techniques. In: 19th world mining congress, New Delhi, Nov, pp 583–595
4. Roy PP (2005) Rock blasting effects and operations. Oxford and IBH Publishing Company Pvt. Ltd., New Delhi (Also published by Taylor & Francis Group plc., U.K.), p 345
5. Verma A, Singh K, Rana S, Keshawanand, Verma HK (2014) Improving productivity & safety by blast design optimization at Ratle hydro electric power project. In: J&K, proceedings fifth Indian rock conference, INDOROCK, 2014, Nov, p 9
6. CSIR-CIMFR (2014) Description of clean technologies and development of environment standards for limestone mining, Technical report No.GAP/007/EMG /CPCB/2007–08 of CSIR-Central Institute of Mining and Fuel Research (CSIR-CIMFR), Sponsored by Central Pollution Control Board (CPCB), New Delhi, (Unpublished), March, p 182
7. Massoud MA, Fayad R, Kamleh R, El-Fadel M (2010) Environmental management system (ISO 14001) certification in developing countries: challenges and implementation strategies. Environ Sci Technol 44(6). American Chemical Society. https://doi.org/10.1021/es902714u,pp.1884-1887
8. EBRD (2012) Mining operations policy—a document of the European Bank for Reconstruction and Development (EBRD); www.ebrd.com › downloads › policies › sector › minin. pp 1–6
9. Saxna NC, Singh G, Ghosh R (2002) Environmental management of mining operation, Scientific Publisher (India) Jodhpur, ISBN 81-7233-296-3, p. 410
10. Lamare RE, Singh OP (2016) Limestone mining and its environmental implications in Meghalaya, India. ENVIS Bull Himalayan Ecol 24:87–100. https://www.researchgate.net/publication/319502640
11. Gibbs MJ, Soyka P, Conneely D (2001) CO_2 emissions from cement production, good practice guidance and uncertainty management in national greenhouse gas inventories, pp 175–182
12. Norgate T, Haque N (2010) Energy and greenhouse gas impacts of mining and mineral processing operations. J Clean Prod 18:266–274
13. Dahlsrud A (2008) How corporate social responsibility is defined: an analysis of 37 definitions. J Corp Soc Responsib Environ Manage 15:1–13 https://doi.org/10.1002/csr.132
14. Moir L (2001) What do we mean by corporate social responsibility? Corp Govern Int J Bus Soc 1(2):16–22. https://doi.org/10.1108/EUM0000000005486
15. Masoud N (2017) How to win the battle of ideas in corporate social responsibility: the international pyramid model of CSR. Int J Corp Soc Responsib 2(4):1–22. https://doi.org/10.1186/s40991-017-0015-y (open access)
16. Chakravorty SL (2001) Artisanal and small–scale mining in India, Country Paper No. 78, Prepared for Mining, Minerals, and Sustainable Development (MMSD) project by International Institute for Environment and Development (IIED), London and World Business Council for Sustainable Development (WBCSD); published as United Nations Documents relating to Mining Industry by UNEP and UNCTAD; Oct 2001, p 81
17. IBM (1982) Monograph on limestone and dolomite, publication cell—technical consultancy, mining research, and publication division, Indian Bureau of Mines, Ministry of Mines, Nagpur, Government of India, Chapter 5, pp. 5–1 to 5–100
18. IBM (2018) Indian minerals yearbook—2018, Part I, general review: status of reconnaissance permits, prospecting licenses and mining leases (57th edn), Indian Bureau of Mines, Ministry of Mines, Nagpur Government of India
19. Hustrulid W, Kuchta M (1995) Open pit mine planning and design: fundamentals, vol. 1, A.A. Balkema, Rotterdam, Netherlands, p 650. ISBN 90-5410-184-9

Chapter 4
Existing Practices in India: Case Studies from Different Geomining Setup

4.1 Introduction

Limestone mining (like any other mining) is generally carried out by creating standard open benches with adequate height and width in the range of 10–20 m. The open-pit limestone mine requires the digging of rocks using excavators or shovels, drilling down the hole, removing over burden with bulldozers, and transporting the limestone via dumpers, tippers, ropeways, or conveyors. At a majority of limestone mines, unit operation of drilling and blasting is the main production operation and crushers are essential components for limestone sizing. In the captive mines of cement plants, *reclaimers* are deployed for spreading the limestone in stockyards.

Most of the limestone mines for cement plant have a *captive* status, meaning that produced limestone is consumed in the plant itself. Such mines are large in numbers and carry special significance with respect to the mineral being described. Because these captive limestone mines are largely owned by the private sector; they are better organized and managed. The Steel Authority of India Limited (SAIL) is a public sector organization of the iron and steel industry that also operates and owns their captive limestone and dolomite mines for raw materials for steel plants. Limestone mines of unorganized and small-scale sectors present an improvised picture compared to captive mines. Similarly, those limestone mines which do not fall into any of these categories, operate similar to any other mine of non-metallic mineral categories, which produce and sell their product as per the market demand and through available marketing networks.

Each limestone mine in India has specific characteristics; therefore, it carries its own identity. In some limestone mines, normal mining conditions are observed, whereas in others, typical field conditions are encountered (e.g., flowing water channel dividing the mining blocks within the lease area). In short, this chapter discusses different case studies encompassing topographies described as karst, normal, coastal, and hilly.

A. K. Soni and P. Nema, *Limestone Mining in India*, Materials Horizons: From Nature to Nanomaterials, https://doi.org/10.1007/978-981-16-3560-1_4

4.2 Case Studies

Five case studies from different geomining setup and operative by different production organizations are presented in this section [1–3, 19–22]. At four places, limestone is currently being mined. The fifth is the Alsindi limestone deposit of the Himalayas, which is a case record for deposit evaluation, where mining is planned in the future [3].

In one case study, the mining is carried out near Saurashtra coast of India in two identical mines, namely the *Narmada* and *Kovaya* limestone mines. These mines are located very close (around 600–800 m) to the Arabian Sea [1, 2]. In contrast, the Manikgarh and Naokari limestone mines both have typical mining case records with identical features [19, 22]. In addition to reviewing their mining operations, an abridged, observation-based environmental management of these mines is also covered in Sect. 4.3. The environmental management scenario of these limestone mines has been influenced by the characteristic features of the region (Sect. 1.7) where the excavation and mining continues. As a result, the industrial perspective assumes great significance.

Case Study 1

Limestone Mining in Karst Topography

(Nongtrai Limestone Mine, Meghalaya)

A *karst* terrain is generally underlain by limestone or dolomite where the topography is chiefly formed by the rock dissolution and may be characterized by sinkholes, sinking streams, closed depressions, subterranean drainage, and caves. The term karst is derived from a Slavic word that means barren, stony ground. It is also the name of a region in Slovenia near the border with Italy that is well known for its sinkholes and springs. Geologists have adopted karst as the term for all such terrain. The term karst describes the whole landscape—not a single sinkhole or spring. A karst landscape is most commonly developed on limestone, but it can also develop on several other types of rocks, such as dolomite, gypsum, and salt. Precipitation infiltrates into the soil and flows into the subsurface from higher elevations and generally towards a stream at a lower elevation. Groundwater flow in a karst aquifer is fast and turbulent due to secondary porosity. Weak acids found naturally in rain and soil water slowly dissolve the tiny fractures in the soluble bedrock, enlarging the joints and bedding planes (https://www.uky.edu/KGS/karst/). This case study is about a limestone mine located in the Meghalaya state of India, representing limestone mining in a region with karst topography. Such mining regions pose challenging problems of different kinds when compared to normal mines or mining regions.

The Nongtrai limestone mine (NLM) is an opencast mine situated in the remotest and far flung border area of Meghalaya owned by M/s Lafarge Umiam Mining Private Limited (LUMPL), which is a Joint venture of Lafarge, Bangladesh, and Umiam Mining Pvt. Ltd, Shillong, India. The Nongtrai mine area lies in the Shella

Confederacy of the East-Khasi Hills of Meghalaya in the NE region between latitude 25° 11′ 16.2″ N–25° 11′ 49″ N and longitude 91° 36′ 52.1″ E–91° 37′ 29.2″ E. The mine is connected by a motorable road from Shillong and Mawsynram. Village Shella and Shella Bazar is within a radius of 2 km from the mine (Fig. 4.1). The nearest big township is Cherrapunjee, which is known as the rainiest place of the world. The village of Nongtrai, after which the mine is named, is about 35 km away (2.5 km aerial distance) from the mine area by road. NLM lies on the bank of the Umiam River, which is situated in the southern slopes of the central plateau of Meghalaya. This position places the mine in the macro catchment of the Umiam River and the microcatchment of the Phalangkaruh River. The Nongtrai limestone mine has a 100 hectare mine lease area allotted, and the entire limestone deposit is of cement grade.

The Nongtrai limestone deposit is a part of the extensive limestone belt on the southern foothill of Meghalaya plateau and is part of the *Cretaceous–Tertiary sequence.* Study of regional geology of the Shella area indicates that it is a part of the extensive *Prang limestone belt* on the southern foothill of the Meghalaya plateau bordering Bangladesh. The limestone deposit of the Nongtrai in the Shella formation (called the *Jaintia Group*) in the west is from the *Eocene age.* These deposits are considered equivalent to the Sylhet limestone formations of the Bengal basin (Table 4.1).

The General elevation or reduced level (RL) of the mining lease area of the NLM limestone mine ranges from 70–190 m. Geologically, the region presents karst topography with the presence of caverns, sinkholes, and solution cavities (like caves) in partly plain and partly rugged hilly terrain (Fig. 4.2). Thus, extensive joints, wide cracks, and fissures are present in all of the existing rock formations. Some of them also are exposed in the mining area also.

Mining and Salient Technical Details

The planning for mining Nongtrai limestone deposits was started in 1999, and the mine was opened in 2003 for development. The commercial production began in 2006. Currently, the mine has plans to increase limestone production capacity from 02 MTPA (million tonnes per annum) to 05 MTPA as per the approved environmental clearance of the government of India. In 2007, this mine faced production restrictions on account of litigation and could not contribute to the state exchequer in terms of royalty. Mining at the Nongtrai limestone mine is carried out through conventional drilling and blasting techniques. Transportation of material from the mine to the limestone sizer and crusher is achieved through hydraulically operated excavators, backhoes, and dumpers (Fig. 4.3). LUMPL excavates the limestone, crushes the same to the desired size, and transports the crushed limestone to the cement plant located in Chhatak, Sunamganj (Bangladesh) through a 17-km-elevated long belt conveyor (LBC), where 7 km lies within the territory of India and remaining in Bangladesh (Fig. 4.4).

Some salient technical details of the Nongtrai limestone mine are as given here.

- Type of mine: Opencast mechanized
- Method of mining: Conventional, drilling, and blasting techniques

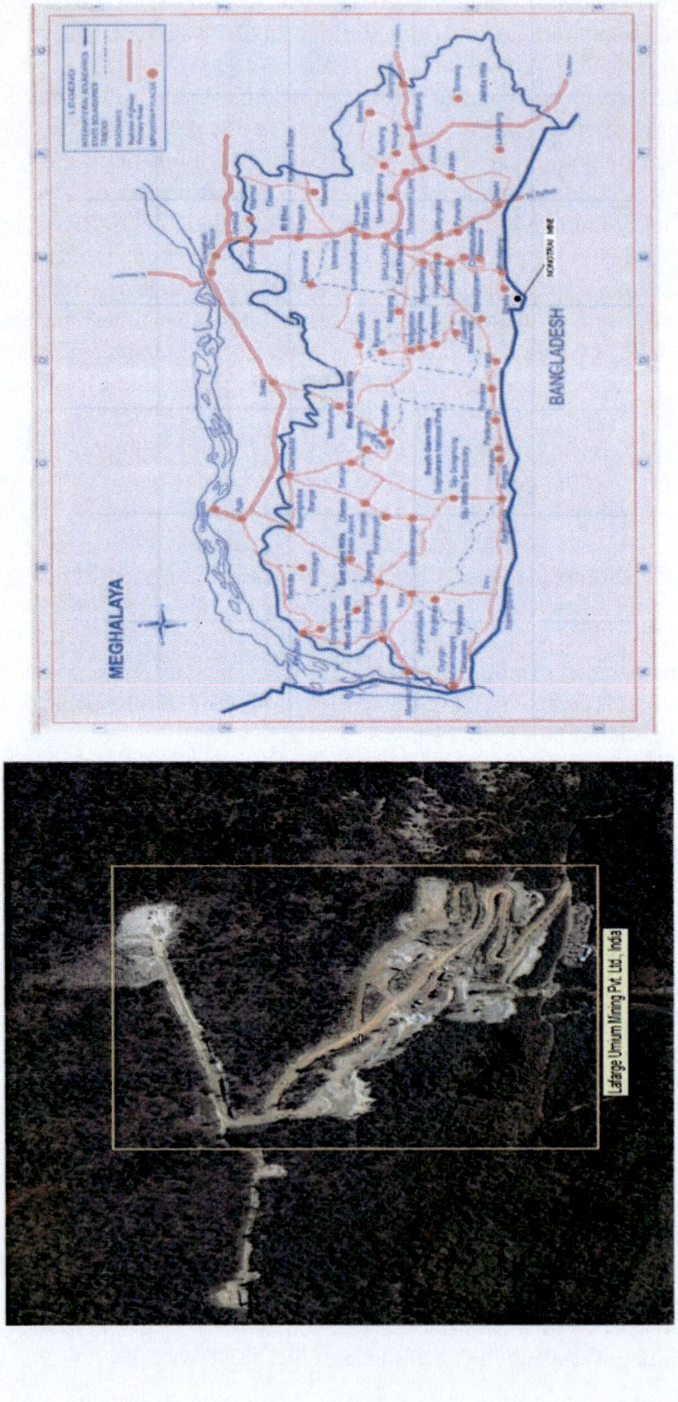

Fig. 4.1 **a** Nongtrai limestone mine as per Google image and **b** location map of mine

Table 4.1 Local geology of study area

Age	Formation	Rock types	Thickness
Oligocene to Upper Eocene	Kopili formation	Light grey to yellowish brown calcareous shale and silt-stone	–
Upper to Middle Eocene	Prang limestone	Hard massive, fossiliferrous grey limestone	100–175 m
Lower Eocene	Narpuh sandstone	Pinkish and brownish white, friable, medium-to-coarse sandstone	20–22 m
Lower Eocene	Umlatdoh limestone	Pinkish white to grey, hard foraminiferal limestone associated with subordinate calcareous sandstone	–

Fig. 4.2 Karst topography and sinkholes in Nongtrai limestone mining area

Fig. 4.3 Mining and ROM transportation at Nongtrai mine

Fig. 4.4 Overland belt conveyor (OLBC) system at Nongtrai limestone mine. *Source* ROM transportation from Meghalaya to Chhatak cement plant, Bangladesh (17 km)

- Geological reserves: 127.36 MT
  (Proved: 87.42 MT, probable: 02.60 MT, possible: 37.34 MT)
- Mineable reserves: 90.02 MT (as on 1/10/2019)
- Life of mine: 18 years at 5 MT per year
- Lease period: 50 years
- Working hours: 2 shifts with 8 h working
- Planned production: 2 MT per annum (7000 tonne/day)
- Number of persons employed: 100 (approximately in mines and ancillary operation)
- Planned bench height: 10 m and bench width: 25 m
- Overall pit slope: 45°
- Ultimate pit depth: Up to 90 m RL
- Explosive used: Class II explosive and ANFO
- ROM handling (for belt conveyor): 800 tonne/h
- Equipment deployed: Hydraulic drills (115 mm diameter); hydraulic excavators (4.5 m^3); backhoe, dumpers (18 tonne); bulldozer; rock breaker; jack hammer; wheel loader; grader, compressor, and other ancillary equipment like a crane, an explosive van, etc.

The *Prang limestone formation* and its outcrops at the Nongtrai mine (in karst topography), being cavernous with pockets, have thicknesses varying from 10–45 m from the surface. The above ground features are characterized by sharp-edged jagged outcrops intersected by numerous fissures, solution cavities, and some large sink-holes. The ridges are dissected by gullies, nala, and rivulets, resulting in the formation of hill slopes with multiple steps. Therefore, the karst topography poses a great risk to the overall mine safety when limestone mining at Nongtrai, warranting several additional precautions. The physiological and geological features of karst topography have conditions where the heavy earth moving machine's operation is unsafe on account of a *solid floor.* This is significant to note here that because of the karst topography the ground (floor) turns very-very uneven/rough with visible natural perforations (Fig. 4.2) making it weak and dangerous. That is also the reason why the word 'solid floor' has been used in previous sentence. At the NLM area, large sinkholes have been encountered while mining, which poses a risk of engulfing the heavy earth moving machines (HEMM). There have been real instance of an entire HEMM being buried in a sinkhole because of soft ground conditions and machine weight. The uncertainty of such conditions that exist pose a hindrance to normal machine operation, making this particular issue extremely important for both machine and workmen safety at this mine.

In this opencast mine, the use of explosives for a *production blast* is quite misleading because of the lack of solid ground. Many times, the mine operator does not get the desired production results because the desired explosive energy is not fully realised. At NLM mine, the need for routine blasting also cannot be performed as it is not possible to make drill holes because of karst ground. This poses another hindrance in normal production.

Moreover, field observation indicates that this area receives higher amounts of rainfall compared to other mining areas. Because of the fractured strata, the rain water percolates downward quickly. These typical features of the surrounding area make the mining risky from the viewpoints of safety and water management, warranting special precautions. Therefore, a tailor-made approach for water management at Nongtrai is required. It was found that since the inception of this mining operation, all major components of the environment (air, water, land, ecology, and society) are being addressed scientifically by the mine operator (i.e., LUMPL).

In general, the mining and ancillary activities are not very difficult, but extra care has to be taken to avoid risk due to the karst topography. Being environmentally sensitive and close to an international border (country boundary), this area requires due regard to the environmental protection and management at the mining site. Its surroundings also have to be kept in accordance with the EMP of mine, which essentially outlines the mitigating plans to combat the mining impacts.

Case Study 2

Limestone Mining in Coastal Area

(Narmada Cement Mine, Gujarat)

UltraTech Cement Limited, a flagship company of the Aditya Birla Group, owns the Narmada Cement-Jafarabad Works (NCJW). This unit of UltraTech was commissioned in 1981 at the village of Babarkot in the Taluka of Jafarabad in the Amreli district of Gujarat. The NCJW has klinkerization and cement manufacturing capacities of 1.5 million tons per annum. The Narmada cement mine (NCM) is the captive limestone mine of NCJW. Today, UltraTech Cement is one of the most dynamically

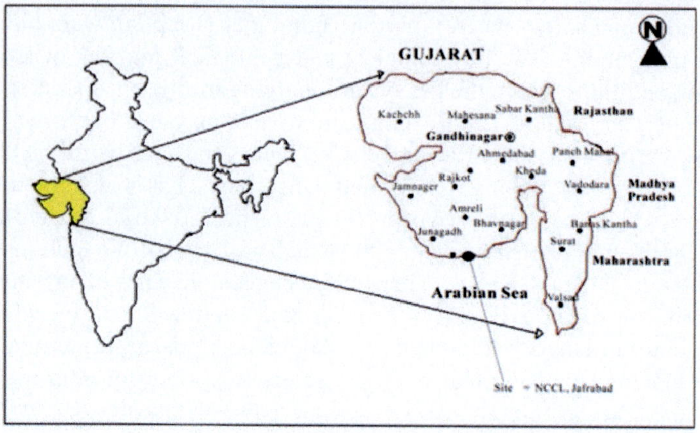

Fig. 4.5 Location map of study area (Amreli district, Gujarat, India)

growing and largest cement producers in India with a 69.3 MTPA installed capacity and having a market share of about 18%.

The lease area of the Narmada cement mine (NCM) lies in the administrative jurisdiction of the Amreli district of the Gujarat state on *the Gujarat coast of India* (Fig. 4.5) and can be traced on the Survey of India toposheet No. 41 P/5 (F42X5, 2010).

The mine is approachable by road (4.5 km) from the town of Jafarabad, which is the nearest urban locality. Distances from the mine site are about 25 km from the town of Rajula, 152 km from Bhavnagar, 215 km from Rajkot, and 95 km from Amreli. The nearest airport is Diu, at a distance of 67 km (approx.) from NCJW, and the nearest railhead is Rajula on, the Bhavnagar-Rajula section of the Western Railway. This case study area has location coordinates at latitude: N 20° 52′ and N 20° 54′ with longitude: E 71° 23′ and E 71° 27′.

The general strike of the limestone deposit at the NCM mine is N 40° E–S 40° W with a dip of 5°–6° towards the east. The NCM mine has a lease area of 565.94 ha (validity up to March of 2030) and mineable reserves of 23.893 million tonne (as on April 1 of 2014). At this mine, limestone excavation (mining) is done above 0 m MSL (mean sea level) in the 261.9 ha *active mining area,* which keeps on changing periodically. Targeted production from the NCM mine (both *limestone* and *marl)* is 2.3 MTPA (million tonne per annum), and it achieves a monthly production of approximately 12,500 tonne.

The NCM is a mechanized mine having two limestone blocks called *east pit* and *north pit* (Fig. 4.6). In 2019, the limestone extraction was being done from these two pits. The mining was carried out using a combination of conventional means as well as surface miners. Excavation of miliolitic limestone leads to the digging of *marl* as the latter is the associated (host) rock of limestone. There is practically no overburden cover over the limestone deposit. A conveyor belt transports the ROM from the crusher to the cement plant. The total quantity of the material dispatched

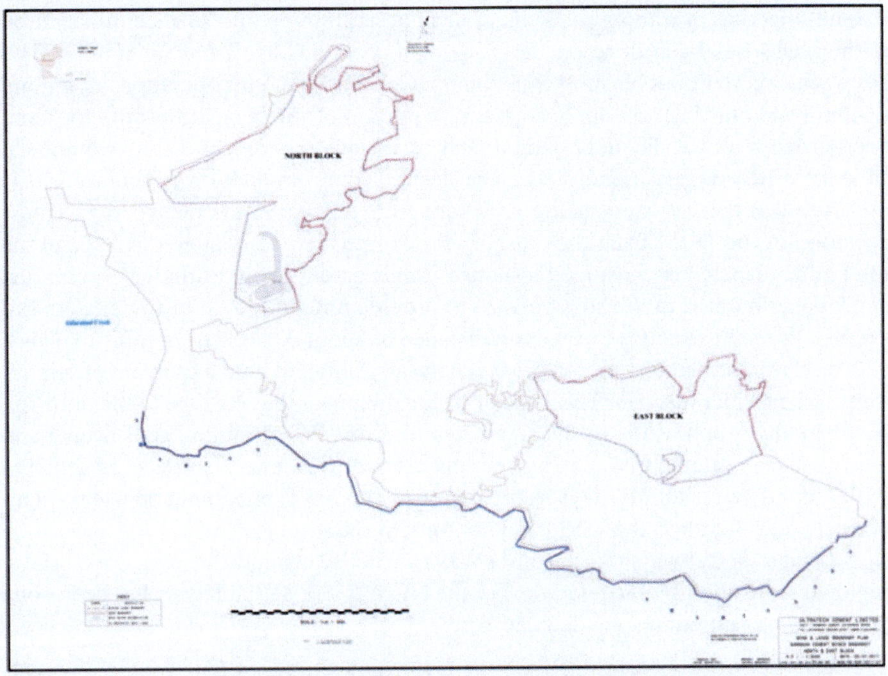

Fig. 4.6 Key map of NCM blocks showing *north pit* and *east pit*

from the mine is consumed in the cement plant, and the target production is achieved conveniently.

The limestone mine of NCJW lies all along the coastline of the Arabian Sea in Gujarat. The NCM mine is a shallow mine producing limestone for its captive use in cement manufacturing and has achieved its rated capacity of production (Table 4.2). Due to restricted depth, the horizontal expansion of the mine is large, and groundwater is not intercepted.

Table 4.2 ROM production at Narmada cement mine

S. No	Year (April–Mar)	Total ROM production (limestone of all grades)
1	2010–2011	1,860,662
2	2011–2012	1,951,286
3	2012–2013	2,020,647
4	2013–2014	1,886,526
5	2014–2015	1,962,182
6	2015–2016	1,676,020
7	2016–2017	1,538,086

The selection of mining methods (conventional mining and surface miner) used at the NCM has been done on the basis of those areas available for mining (i.e., areas with and without blasting constraints). Conventional unit operations of mining (drilling and blasting) are applied on a hard portion of the deposit for both rock and overburden removal. By deploying a fleet of various equipment, such as dumpers, shovels, drills, dozers, and graders, the desired mine production is achieved. It is observed that the limestone is not very hard from an excavation perspective. Transportation of the ROM from the crusher is accomplished by conveyor belts, and the total mine dispatch is consumed in-house at the cement plant. Sufficient yet limited stock is maintained in the storage yard to provide a regular feed of raw material to the NCJW plant, which is located at a distance of about 0.8 km (from mine to plant).

The Narmada cement mine (NCM) is naturally inflicted with a problem of salinity ingress (sea water intrusion) because of its location near the sea. Due to the intrusion problem, the groundwater of the area becomes salty. When mining goes deeper and water table is intercepted, nearby regions around the mine also become inflicted with groundwater salinity. This saltwater intrusion (SWI) phenomenon occurs at all coastal areas—around the world and not only in India.

Seawater or saltwater intrusion (SWI) can be determined in two ways: (1) at regional scale and (2) at local scale. For the NCM mine, SWI analysis has been done [1, 2] and those findings are detailed here:

1. The Narmada cement mine and its host rocks are limestone, which is a sedimentary formation with adequate porosity for groundwater flow. Such formations cause continuous dissolution of calcium in water, resulting in higher total dissolved solid (TDS) in water. Because of *continuous dissolution* taking place at both shallow and deep levels, such formations always have higher TDS levels.
2. This limestone mining area forms part of the limestone belt of the Saurashtra region along the coast of the Gulf of Khambat. Major parts of the mine lease area of NCJW lie near the shoreline of the Arabian Sea.
3. In general, the physiography of this region resembles any coastal region. Land-derived material from volcanic activities creep in as impurities in the limestone deposit, as well as various types of soil, alluvium, windblown sand, marine fluvio-mud deposit tidal flats, and shell and shingle deposits on the shore area that commonly occur. The highest elevation in the mine lease area is 35.25 m AMSL (above mean sea level) towards the north-west.
4. Miliolitic limestone is the major water-bearing formation in this study area. Considering the watershed concept of resource planning, the NCM area is bounded by the ridge on the west (W), the Dhatarwadi River on the east (E), and the Arabian Sea on the south (S). The watershed boundaries are spread between latitudes N 20° 50′–21.00° and longitudes E 71° 29′–71° 23′, and the mine lease area is surrounded by a number of villages, namely Babarkot, Varahswarup, Bhakodar, Kovaya, and Vand (Fig. 4.7).
5. Younger geological formations made up of windblown sand constitute the overburden with an average thickness of 01 m. Occupying the flatter areas, the thickness of overburden is comparatively more towards Kovaya and villages,

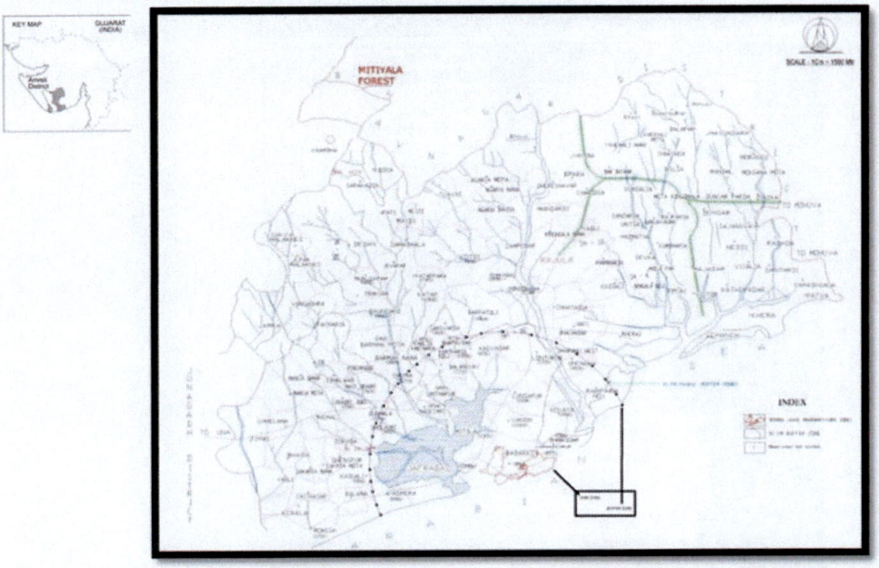

Fig. 4.7 Core zone and buffer zone of NCM

while the thickness decreases towards the village of Babarkot, which is on comparatively higher ground. Just below the windblown sand, miliolitic limestone occurs as compact and low-grade limestone of varying thickness with the average being approximately 25 m. At a greater depth, there is a general deterioration in the quality of limestone accompanied by soft, friable layers and lenses of low-grade limestone. This limestone is buff, yellow, or pinkish in colour, while the weathered surfaces show a sooty to brownish appearance. The main constituent of this limestone is the calcic shell fragments of the miliolitic variety. These limestone deposits show enrichment in calcium content at the upper layers, probably formed due to the leaching out of silica with the calcium enrichment caused by capillary action and redeposition in crevices. This limestone is compact, hard, and boulder-like in nature at the study area at many places. The average quality of limestone shows CaO: 40–48%; SiO_2: 4–7.5%; Al_2O_3: 1.5–3%; and Fe_2O_3: 1%–2.5%, which is cement grade.

6. A study of the region and mine area based on a *resistivity survey, water quality analysis,* and *ground water modelling* (completed using SEAWAT2000 software) has shown that the *seawater intrusion* (SWI) is present in both east and north pits of the NCJW limestone mine. The extent of seawater intrusion that covers the lease area also has been assessed on the basis of groundwater modelling and simulation results for future scenarios. The modelling results showed that saltwater intrusion (*interface line*) has extended up to 1.2 km inside the coastline and towards the mainland (Figs. 4.8 and 4.9). From this analysis, it also has been concluded that the saltwater has intruded very slowly over the

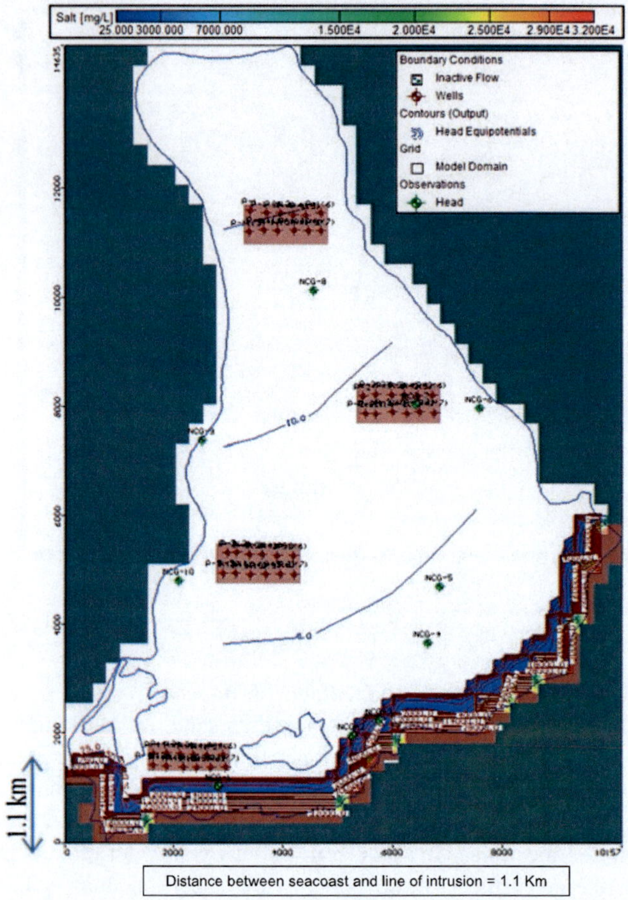

Comp/Art: Distance box is part of art (like a legend) not part of caption

Fig. 4.8 Results of a five-year simulation period (SEAWAT 2000 output)

years. NCJW mining pits will have vulnerability from the SWI now and in the future (i.e., 2017 and beyond) [2].

CIMFR has observed that pit mining at Narmada cement mine for limestone extraction can be done easily with SWI present up to (–)4 m MRL to (–)8 m MRL safely and conveniently [2]. It also has been observed that a *seawater–freshwater interface* can be kept under control by adopting water management measures, such as less groundwater draft and adequate groundwater recharge. Suitable *water management* measures play a positive role in SWI containment.

Further, another large limestone mine, Kovaya limestone mine, of Gujarat Cement Works (GCW), which is owned by M/s UltraTech Cement (same company as that of NCJW), is also operating near the NCM mine. The Kovaya limestone mine is

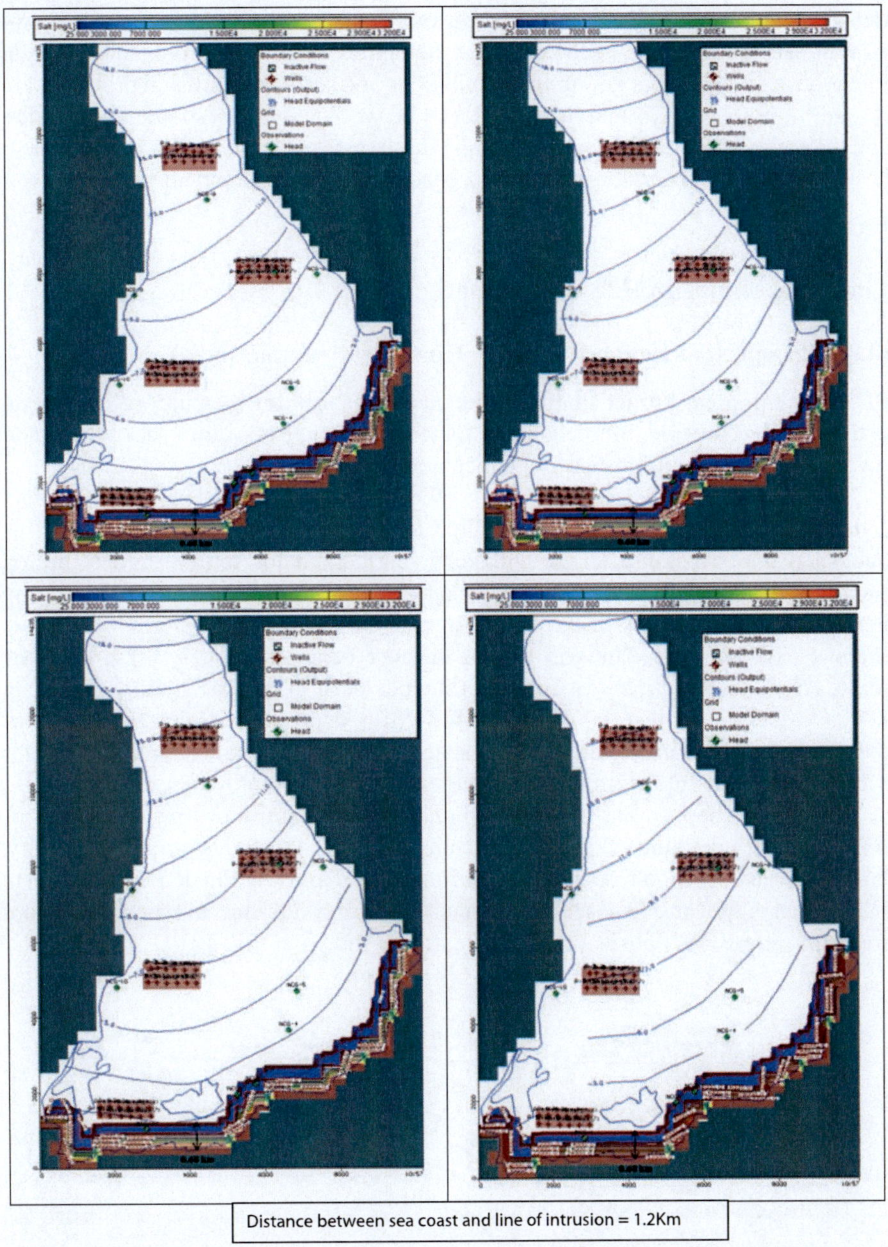

Fig. 4.9 Results of 20-year simulation period (SEAWAT 2000 output) (Clockwise: 0 m; −4 m; −8 m and −12 m depth)

located near Kovaya village—along the same Gujarat coast—and lies at about a 8–10 km distance from the NCM mine [1]. Scientific studies at both of these coastal mines have revealed that geo-mining conditions and seawater intrusion problems are present in both pits. To maintain a balance between mineral resource exploitation and the environment, the company owning these mines (i.e., M/s UltraTech Cement) has adopted an appropriate scientific approach for SWI containment.

Case Study 3

Limestone Mining In Hilly Topography

(Alsindi Deposit of Indian Himalaya, Himachal Pradesh, India)

Plain and hill areas attract sharply different approaches for the purpose of mineral extraction that include mine planning, mine designing, excavation scheduling, and investigation required for their implementation in practices.

In a hilly stretch in Karsog Valley in the Mandi district of Himachal Pradesh, India, there lies a rich cement grade limestone deposit near Alsindi village. This deposit (called Alsindi after the village) is earmarked for setting up a 3 million tonne per annum dry process cement plant. The proposed cement manufacturing plant will be on the bank of the Sutlej River and in immediate visibility from state highway SH-13. This plant will lie at a distance of approximately 8 km from the mine. Accordingly, a lease of about 800 ha has been granted to Lafarge, India, in 2010 for detail investigation. Alsindi deposit (hilly deposit) was selected as a part of the state industrial development by the state geological department and mining was scheduled in near future when planning of state development will be taken up.

The Alsindi deposit has the required quality and quantity of limestone reserves for setting up a medium- to large-scale cement plant with captive mining feasibility, but the terrain is difficult. As the Alsindi limestone deposit is situated in lofty Indian Himalayan region, it is exposed to inclement weather conditions. In Himalaya, *domal* forms of limestone deposits occur frequently, as shown in Fig. 4.10.

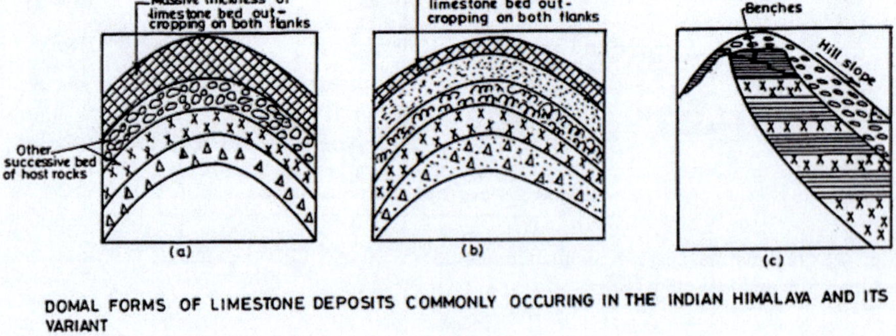

Fig. 4.10 Limestone deposits of the Himalayan region

Some other variants of limestone deposits that have been reported in the Himalayan region include the following.

1. Limestone beds with massive thickness and outcropping on both flanks
2. Limestone beds with limited thickness and outcropping on both flanks
3. Limestone beds dipping along one side of the hill slope (with both massive and limited thickness). The limestone beds dipping along the slope may have lie along the slope or against the slope. In the latter case (i.e., against the slope), the overburden thickness increases progressively.

Deposit forms of Alsindi, as confirmed by an L section study of this deposit, fall into category 1. The Alsindi deposit has a massive thickness with outcropping on both flanks where the entire hill is composed of limestone mineral. The salient features of the area, its mining, and other geological details are described next.

The limestone-bearing area around Alsindi, where an open pit mine has been planned, is situated at a distance of about 70 km north of Shimla and lies in the Mandi district of Himachal Pradesh (Fig. 4.11). The site can be approached from Shimla by the state highway No. SH-13, which is an all-weather metalled road that remains open for traffic throughout the year. The town nearest the mine is Sunni (near Tattapani) located at a direct distance of about 10 km south on the south bank of the Sutlej River.

Discussed limestone deposit falls in Survey of India topographical sheet No. 53E/3 (unrestricted) and is located between latitude 31° 17′ 10″ N and longitude 77° 05′ 77″ E. The entire 800 ha mine area has an average altitude ranging from 1100 to 1800 m. Nearest villages in the mining area are Badgeogshao, Talain, and Udayanal (Fig. 4.12), all located on various adjacent hills in the rugged terrain.

On plan, Alsindi limestone deposit has length of 8000 m and width of 1050 m approximately with a mineral depth of over 200 m from the surface. The mine site is characterized by hilly topography and has all of the essential features of rugged Himalayan terrain (i.e., ridges and valleys). In and around the deposit, the average altitude varies from place to place.

The mineable reserve of the Alsindi deposit is 150 million tonne, which is sufficient for approximately 45 years of 2–3 MT cement production (@ 8000 TPD or 3 million tonne of annual limestone production). With such sizable limestone reserves, a large surface miner can be planned to feed raw material to a cement plant.

The limestone deposit at Alsindi belongs to the Sorgharwari formation of *Shali Group* of the Proterozoic age. Within the Shali Group, the Sorgharwari formation is underlain by the Khatpool formation and overlain by the Tattapani formation. This deposit has a general strike towards NW–SE and a dip of about 23°–72° towards the NE. The limestone bed has two distinct sequences from the lower pink to purple limestone and the upper grey limestone. The pink-to-purple limestone is very fine textured, creamy, well laminated, and banded. It is also fine-grained, dense, and homogeneous, and it exhibits conchoidal to subconchoidal fractures. It contains some green shale partings. The grey limestone interbeds with the pink limestone before grading into Tattapani formation. The observed lithologic (litho-unit) sequences are given in the following table:

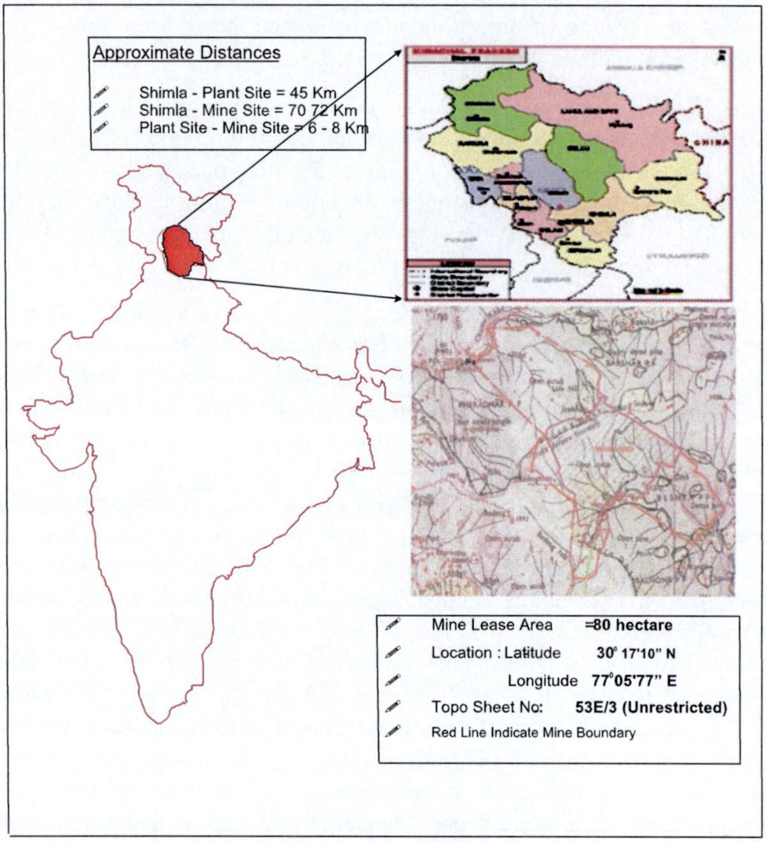

Fig. 4.11 Location map of the Alsindi deposit

Soil	Greater part of lease holding area is covered by a thin layer of sticky black 'cotton' soil derived primarily from the weathering of the basaltic rock occurring in the elevated hills. The thickness of the soil ranges from 0.20 to 3 m
Limestone: Sorgharwari formation	The Sorgharwari member consists of pink-to-purple limestone as well as dolomite, which is exposed near the village of Alsindi. This limestone is present in the entire area and also outcrops at some places
Dolomite: Khatpul formation	The Khatpul formation mainly consists of dolomite. It occurs in a limb of overturned anticline form of fold and is exposed all along the western part of the area. The dolomite is massive, hard, and compact and shows typical weathering

Acute angled slopes are prone to rockslides; hence, they will require suitable scientific planning and preventive measures. This mining site will always face slope

Fig. 4.12 Alsindi deposit location as observed from Google Maps. *Courtesy*: Google, India

stability issues because the area is a part of hilly Himalayan region, and the rocks of the terrain have undergone a series of folding and faulting as a result of tectonic activities. During mining or actual excavation, the presence of *shear zones* and *overturned folds*, with an axis of the fold trending along strike direction (i.e., NW–SE), may be encountered or exposed because geological and structural conditions are frequently varying in the Himalayas.

An assessment of technical feasibility for mining of Alsindi limestone deposit of Himalaya was made in 2012, and results were given in a technical report [3]. Some perspective views of the mining pits and the study area from different angle and locations are as depicted in Fig. 4.13a–c. Based on the scientific assessment of the area, it is revealed that mining can be planned in two open pits as the *lower pit* and the *upper pit*. The Jankhuni Ridge is in close vicinity of these pits and also may be a potential limestone source for extraction in the future, as the number of hills near this mining site are composed of limestone.

At this deposit site, limestone and dolomite both maybe struck while digging. From a mining point of view, Alsindi has the form of an excellent massive deposit, but it lies within a difficult terrain with unsteady climatic conditions. Its analysis

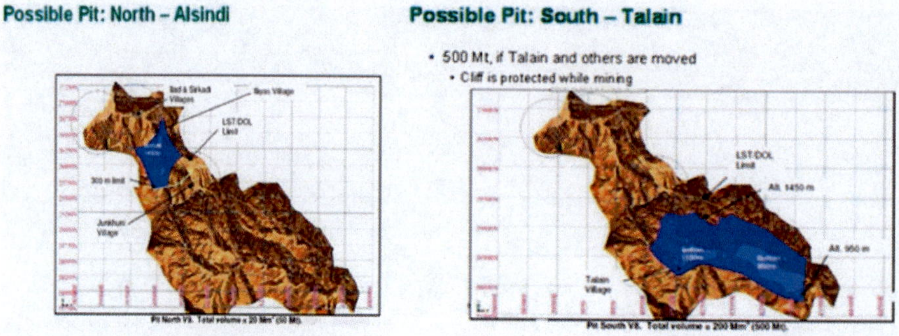

(a) 3D view of north mining pit **(b)** 3D view of north mining pit

Fig. 4.13 Views of mining pits of Alsindi deposit (H.P., India)

shows that *opencast stripping* is a feasible option for mining the limestone from here. However, to protect and preserve the serene Himalayan environment (an eco-sensitive region), utmost care needs to be taken. This evaluation study [3] also concluded that limestone exploitation should be resorted to as the last option if needed for regional development.

Case Study 4

Limestone Mining in Normal Conditions

(Partipura Limestone Mine)

(*Courtesy*: The India Cements Limited, Banswara Works, Rajasthan)

The India Cements Ltd., Banswara works have a 1.8 million ton capacity cement manufacturing unit at Jhalo-Ka-Garha in the Banswara district of Rajasthan. Their captive limestone mine, Partipura Limestone Mine (PLM), is named after the local village and has a mine lease area of 65.82 hectare allotted for open pit limestone mining. The Partipura limestone mine and the core mining area are located in Garhi Tehsil of Banswara district, Rajasthan (Fig. 4.14a). The mine is situated in close vicinity of the cement plant (Fig. 4.14b) and has an environmental clearance (EC) to produce 1.872 million ton (MT) of limestone per annum. The Partipura quarry is presently working up to the seventh bench level (i.e., 42 m below the ground level). With the progression of the mine's age, such surface mines are further deepened for limestone extraction. This mine has already encountered groundwater, and mining is being continued with the CGWA permission by dewatering the mine pit.

The Partipura mine area is a hard rock area consisting of undulating terrain that is largely plain. Limestone of the Partipura area is of cement grade. There is a surface outcrop at very few places. The thickness of the soil cover varies from 0.5 to 2.0 m. Mechanized limestone production and ROM handling by automated means have been in practice at this mine.

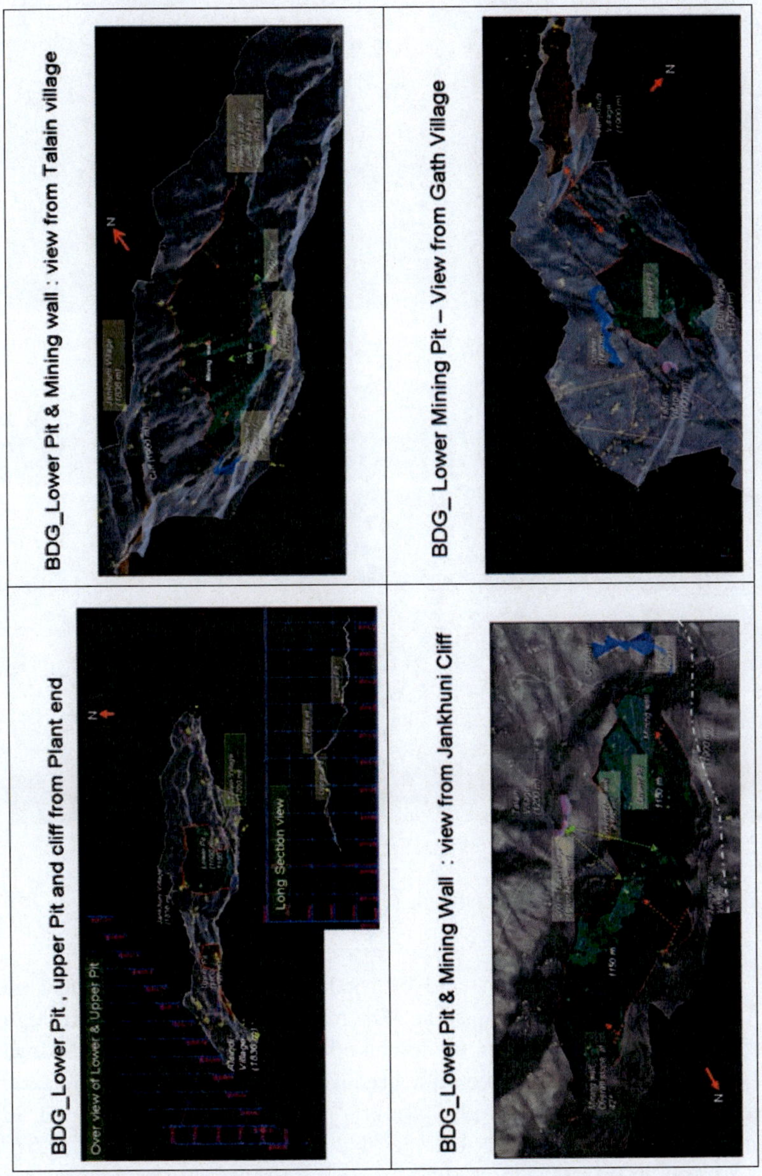

(c) Views mining pits as seen from different angles

Fig. 4.13 (continued)

(a) A view of captive Partipura limestone mine (PLM) of India Cement Ltd., Rajasthan

(b) The India Cement Ltd. plant, Banswara district, Rajasthan

Fig. 4.14 Partipura limestone mine (PLM) and India Cement Ltd., Rajasthan

The PLM study area is well connected by road from the towns of Banswara (24 km) and Dungarpur (85 km) of Rajasthan. The mine can be approached through the interstate highway passing through the towns of Ratlam and Indore of Madhya Pradesh via Banswara (Fig. 4.15a). The mine area (i.e., core zone, buffer zone, and adjacent area) lies in the Survey of India topographical sheet Nos. 46 I/2 and 46 I/6. The exact mine site location falls between N23° 36′ 10.93″: N2° 36′ 53.57″ latitude and E74° 14′ 25.06″: E7° 15′ 12.95″ longitude. However, surrounding area of the mine consists of a ten kilometre radius falling between 74° 8′ 54.21″ E and 74° 20′ 39.25″ E longitude and 23° 31′ 6.16″ N–23° 41′ 55.96″ N latitude. Latitude and longitude of the mining lease area corner points are as shown in Fig. 4.15b.

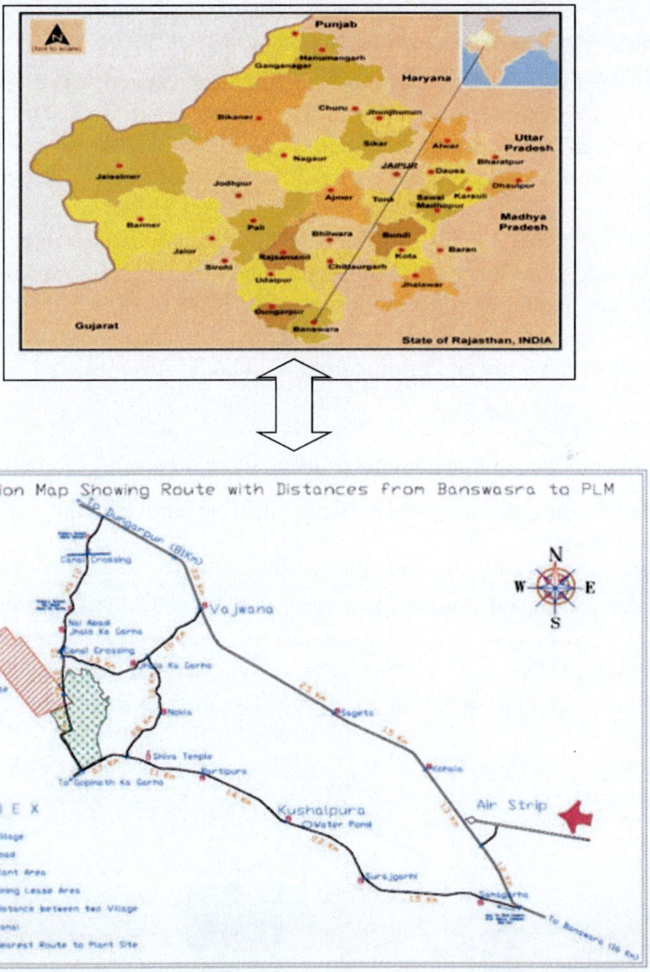

(a) Location map showing route with distances from Banswara to PLM

Reference Points	Latitude	Longitude
Q	N23°36.304'	E74°15.219'
P	N23°36.185'	E74°14.929'
B1	N23°36.805'	E74°14.760'
B2	N23°36.780'	E74°14.728'
B3	N23°36.894'	E74°14.623'
C	N23°36.738'	E74°14.421'

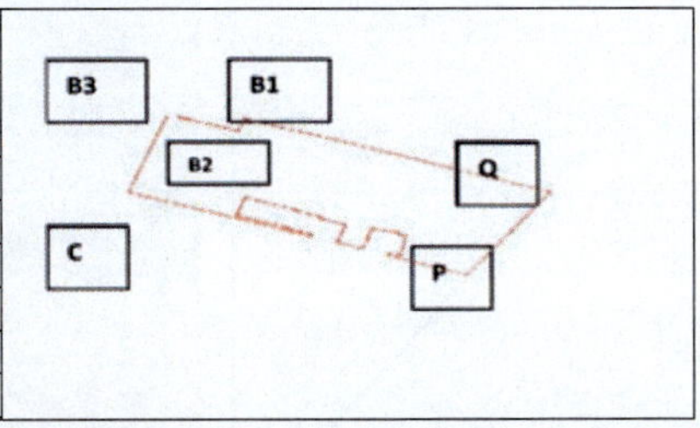

(b) Coordinates (latitude andlongitude) of corner points of mining lease area

Fig. 4.15 Partipura limestone mine (PLM) location data

Mining at PLM is carried out using conventional drilling and blasting techniques (Fig. 4.16). The ROM transportation to the crusher installation uses dumpers, hydraulic operated trucks, and tippers of medium capacity (Fig. 4.17). A crusher is used for sizing limestone fed into the cement plant. The active working zone is concentrated on the south side of the leased area and advances in the northern direction. The PLM boundary lies in close proximity of the cement plant, so the transportation distance of raw material from mine to plant is very little (optimum). There are two active dump sites proposed to accommodate an inter burden of mine materials. Adequate space is provided for the crushing plant, mine roads, and mine infrastructure (including office buildings). The PLM is a newly developed mine, and additional developmental works are envisioned for capacity expansion in the future. Key details of the mine and mining operation are as given next.

Mining at PLM

Type of mine: Opencast Mechanized (Mine lease area = 65.82 ha)

1. Method of mining: Conventional; drilling and blasting technique (2 shift working)
2. Mineable pits: One single pit
3. Total geological reserves and resource: 80.429 MT; Proved reserves (Code 111): 26.114 MT; Probable reserves: 11.47 MMT; Remaining resource: 42.846 MMT

   (*Source*: Data as on 01.04.2018; MMT = million metric ton)

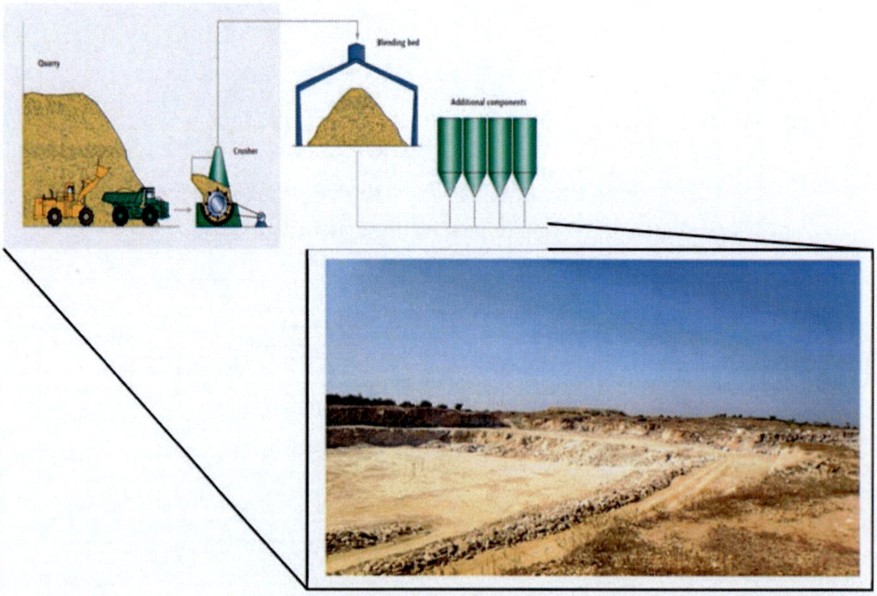

Fig. 4.16 Conventional mining at Partipura limestone mine

Fig. 4.17 ROM transportation at the Partipura mine

4. Limestone deposit strike length: 1.5 km
5. Life of mine: 24.21 years (at current rate of production about 340 working days per year)
6. Present production rate: 6000 tons/day
7. Bench height: Bench width = 8–10 m
8. Bench alignment: NE–SW direction across the strike
9. Highest upper level: 200 m MRL
10. Lowest level (proposed): 130 m MRL
11. Working levels (as on 01.04.2019)—1st bench 186 mRL and above; 2nd bench at 180–186 mRL; 3rd at 176–180 mRL; 4th bench at 170–176 mRL, 5th bench at 162–170 mRL; 6th bench at 154–162 mRL; and 7th bench at 148–152mRL
12. Ultimate working depth: 70 m bgl
13. Limestone production (current and planned) at 6000 tpd
14. Proposed expansion (if any) = NIL
15. Explosive used: Class II explosive and ANFO
16. ROM transportation: Mine face to crusher by tippers; crusher to plant by conveyor
17. Equipment deployed: Drills (4″–4.5″ diameter) = 02 Nos.; shovels or excavator (2.1 m^3) = 02 Nos.; tippers (21 tonnes) = 12 Nos; bulldozer (D6R) = 01 No.; rock breaker with excavator (80 tons/h) = 01 No.; and backhoe with loader = 01 No.
18. Other ancillary equipment: Compressors, crane, explosive van, etc.

Table 4.3 Production of limestone at PLM

Year	Total limestone production (in tonne)	Total interburden handling (in tonne)	Mining RLs (lowest RL of the mine) (m RL)
2015–2016	1,373,477.85	9119.00	170
2016–2017	1,269,436.34	44,015.00	162
2017–2018	1,252,925.23	26,628.00	157
2018–2019	1,328,920.43	95,970.50	157
2019–2020 (until Dec. 2019)	904,128.41	115,271.80	152

Source The India Cements Ltd

A number of villages lie in the buffer zone of this study area, and the nearest village to the mine is Nayatalao at a distance of only 1.06 km (aerial distance). Except for the cement plant premises, all other local villages are at a distance of more than 2 km from the mine. The industrial operation of plant and actual mine operation at this industrial site was started in March of 2010 and June of 2010, respectively.

Limestone production figures are shown in Table 4.3. Limestone production at this mine is almost at the same level, since the Partipura mine had achieved its full capacity production in 2012–2013. As per the mine records, in 2012–2013, PLM produced approximately 1,278,550 tonne. Since then, the mine production and plant requirements both remained almost stagnant. The Partipura mine area limestone reserve is sufficient enough to meet the future requirements of its captive cement plant for another two decades. The mining (quarrying) and excavation procedure is normal and easy.

Case Study 5

Limestone Mining in Typical Field Condition

(Naokari Mine and Manikgarh Mine, Maharashtra)*

This is a case study of two mines—the Naokari limestone mine and the Manikgarh limestone mine both located in Maharashtra state of India. Both mines are typical yet different from that of the normal mine and mining area in the sense that both of these mine leases are traversed by a seasonal flowing surface water channel locally called a *nala*. Under such conditions, and particularly during monsoon period, the water quantity and water quality in the nala, including its management, pose a great challenge. On one hand, the statuary regulations do not permit any disturbances to the nala, and on the other hand, production planning and scheduling becomes difficult. Due to the presence of the nala within the lease area, a significant amount of limestone reserve is blocked, requiring an engineering solution. Also, water pollution becomes a cause of concern in these mines, so pollution abatement measures are also required.

Based on geohydrological study and geochemical analysis, the evaluation of ground and surface water in and around the mining lease area has been assessed with the feasibility of a *nala diversion* from within the lease area to the outside. The information gained from the study is listed here:

- Ease in overall planning of mine
- Better mine design and simplified statuary compliance
- Good mineral conservation practice in terms of extracting locked-in limestone all along the nala stretch and in particular below the nala bed
- Improvement in the mine production and productivity
- Better use of water through pollution control and water quality evaluation.

*For the complete case study refer to Annexure C.

Introduction

Two identical and comparable limestone mining areas of India are the *Naokari limestone mine* (NLM) and *Manikgarh cement limestone mine* (MLM), which are traversed by a flowing surface water channel in the middle of the lease area and have been assessed in this case study. The water channel, seasonal in nature, is referred locally as *BOP Nala* (in the NLM area) and *Amal Nala* (in the MLM area). The water channel divides the mineral property in two separate mineable blocks lying within the mine lease area. Thus, both the NLM and MLM mines face difficulty in terms of normal mining, transportation, and scheduling. Hence, this case study analysis is important from two angles: (1) water pollution and (2) effective utilization of water (both mine water and nala water). It has been observed that the area is an agricultural belt where the need of water for irrigation is high. The BOP Nala and Amal Nala water is valuable for the people residing all along them, as they make use of the water channel in significant quantities.

Human intervention from industrial activity, such as mineral extraction, impacts both the quality and quantity of surface and groundwater resources in many parts of the world [4–8]. In this context, numerous studies on the interrelationship between water quality and mining, water quantity and geology, etc. have been carried out [9–13] and have established that the mining activity can pollute water resources and cause noticeable changes in the water regime [14].

To recognize the impact of mining on water regime, one should consider the mineral type being extracted and the kind of pollution that may be caused at that specific site (e.g., in the case of coal mining, acid mine drainage (AMD), or acid rock drainage (ARD)). In the case of metallic mineral mining, consider the heavy metal content in water. In the case of limestone mining, it is the water hardness, total dissolved solid (TDS) concentration, and fluoride contamination that are the problems. Although limestone bearing mine areas do not pose severe pollution problems, it has been observed that mine water has great variations in the concentrations

Table 4.4 Locations and particular details of NLM and MLM

S. No	Particulars	MLM	NLM
1	Location details (Taluka/district and state)	Korpana Taluka, Chandrapur district, Maharashtra state, India	Korpana Taluka, Chandrapur district, Maharashtra state, India
2	Survey of India toposheet No	56 M/1 and 56 M/2	56 M/1
3	Latitudes	19° 38′ 30″ N–19° 50′ 30″ N	19° 47′ 0″ N–19° 48′ 0″ N
4	Longitudes	79° 04′ 00″ E–79° 10′ 00″ E	79° 07′ 30″ E–79° 10′ 0″ E
5	Distance from the nearest town of Rajura, Chandrapur district	22 km east	30 km east

of various chemical constituents. In and around the study area (i.e., Gadchandur, in the Chandrapur district in Maharashtra, India), a number of cement factories are currently operating and the industrial water demand from such operations (particularly groundwater) is quite high.

About the Area

Location details and particulars of the selected mines of NLM and MLM being investigated in this case study) are given in Table 4.4.

These two mine areas are located adjacent to one another and geologically belong to the same limestone sequence of the Penganga Group (Proterozoic age) overlain by the lower Gondwana sediments (i.e., Talchir, Barakar, and Kamthi formations). Deccan trap basalts, alluvial and laterite are overlying the Penganaga limestone group. The general geological setting as encountered in the study area is shown in Fig. 4.18, and the lithological succession is summarized in Table 4.5. The lease boundaries of the NLM and MLM mining areas along with village names and locations are shown in Fig. 4.19a. The topography of these two mine areas is partly hilly and partly plain. Both of these mines exist in continuation with one another and have the same watershed areas.

NLM is an opencast mechanized mine with two limestone blocks, referred as blocks A and B, that are working blocks for day-to-day production of limestone. The A and B mining blocks are separated from each other by a nala (stream) called the *BOP Nala,* cutting the deposit in two parts and flowing along a S–N direction within the mining lease area (10.3 km^2). Similarly, the MLM mine area on the southern side of the watershed has the *Amal Nala (*also called Nala Vaagu*)* flowing in a S–N direction midway across the mining lease area. Amal Nala divides the mining of ore into A, B, and C blocks and D, E, and F blocks, and it flows in the mining lease area (4.93 km^2), cutting across the deposit (Fig. 4.19b). These water channels, although categorized as seasonal, have water flowing for three-quarters of the year. Compared to BOP Nala of the Naokari mine, Amal Nala contains more water as it has several feeder springs too. The water of Amal Nala is collected downstream in the *Amal*

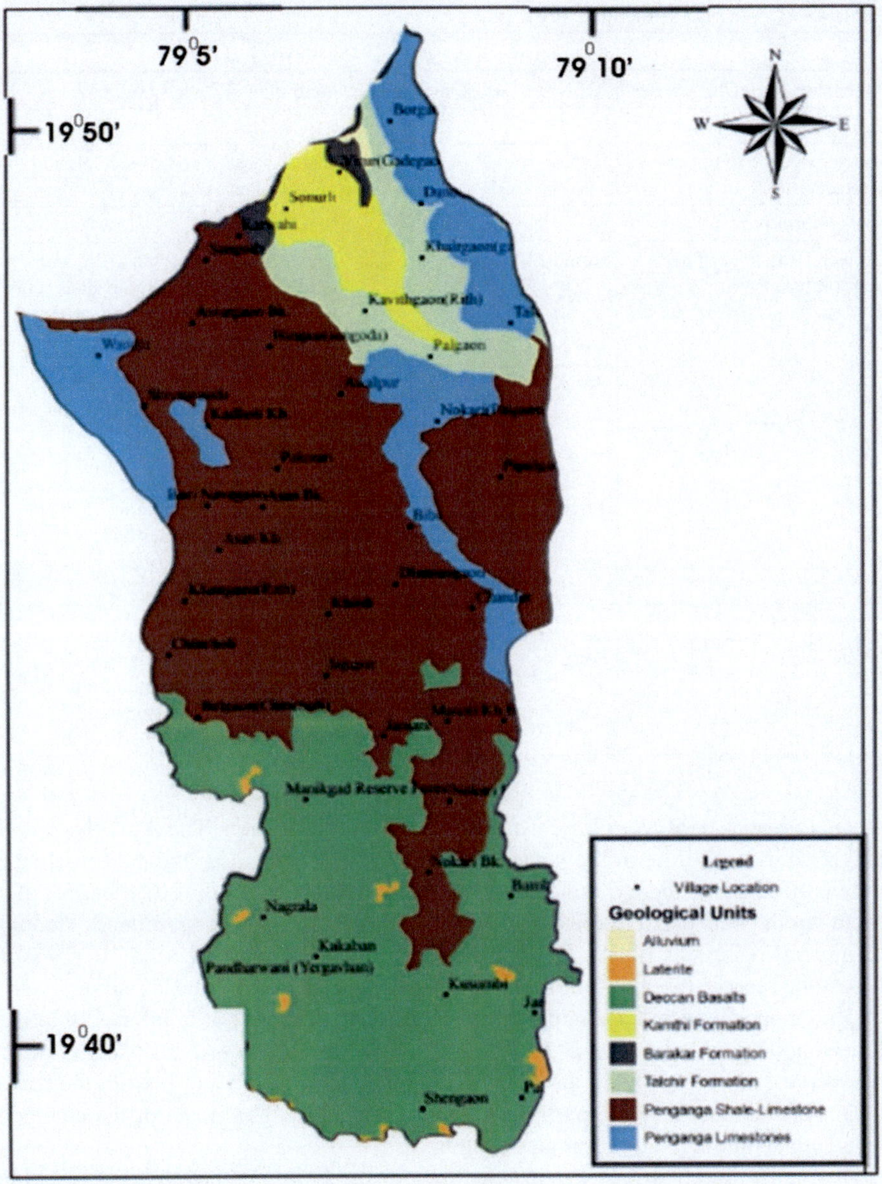

Fig. 4.18 Geological map of the NLM and MLM area

Table 4.5 Geological setting of study area

Age	Geological succession	Lithology
Recent to somewhat earlier	Alluvium, soils, laterite	Sand-clay-silt, soils, and laterite
Unconformity		
Lower Eocene to upper cretaceous	Deccan trap	Basalts, weathered, vesicular, and massive basalt
Unconformity		
Lower Triassic to upper carboniferous	Kamthi formation	Clay, shale, sandstones with coal reddish brown sandstone, shale, clay with coal
	Unconformity	
	Barakar formation	Light grey to white feldspathic sandstones, carbonaceous shales, coal seam, and clay
	Unconformity	
	Talchir formation	Greenish to dark olive green-coloured shales and coarse-grained sandstone with boulder beds
Unconformity		
Proterozoic	Penganga group	Shale, sandstones, flaggy, and massive limestone with alternate bands of shale limestone and sandstones of variegated colours

Nala Dam and *Amal Nala Reservoir* and is used for irrigation. The Amal Nala and irrigation dam both lie in the same catchment area. To maintain the nala existence within the lease boundary, mine management has to leave a natural rock barrier of a 60 m width. This barrier consists of limestone minerals; thus, a significant amount of mineral reserve is blocked. According to a rough estimate, around 10 years of limestone reserve (about 50 million tonne in each mine) is blocked.

Based on scientific facts and geohydrological studies [15], it is inferred that the limestone locked below the BOP Nala only (not Amal Nala or Nala Vagu) can be recovered fully by applying an engineering approach of diverting it from the lease area. This diversion can be partial or full, depending on the impact of the changed nala course on the natural drainage system.

The field investigations carried out previously showed the following possible engineering solutions.

1. Diversion of the total nala stretch within the lease area.
2. Partial diversion of nala stretch upstream of the lease area when it is seasonal.
3. Diversion of nala (total) on upstream stretch by constructing a barrier (i.e., dam or some civil engineering structure).

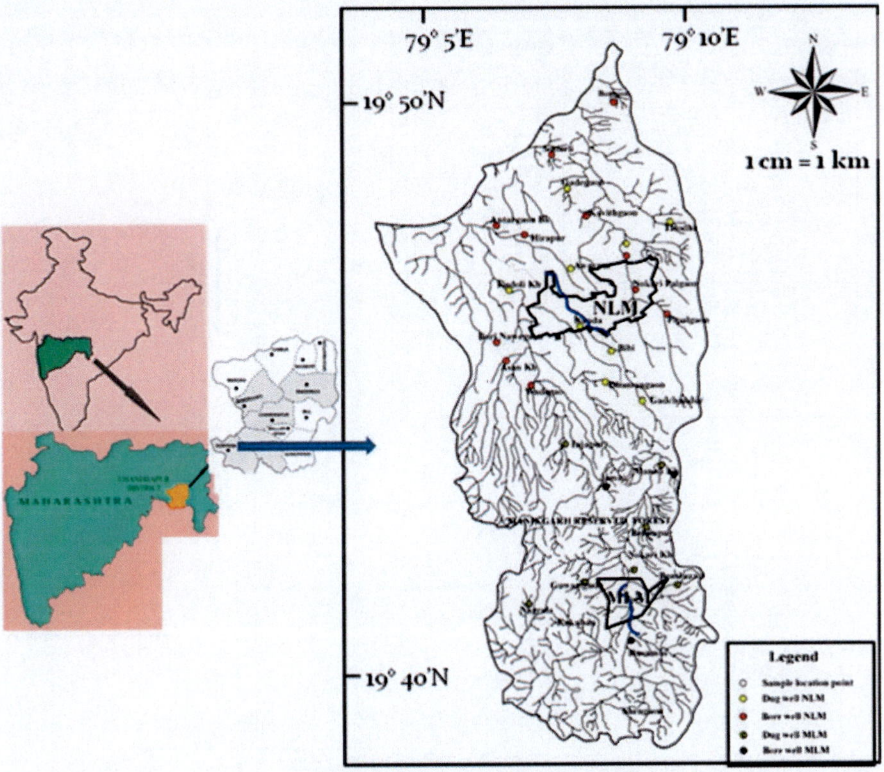

(a) Site map and study area

Fig. 4.19 Naokari limestone (NLM) and Manikgarh cement limestone (MLM) mining areas

The solution mentioned under the second point (i.e., partial diversion) is assessed as an easy, economical, most suitable, and implementable solution. When this analysis is applied to the Amal Nala, it is emerged that the Amal Nala is a feeder water channel (nala) for both the Amal Nala Dam and Reservoir) on the downstream side, and its diversion will cause major disturbances. Therefore, Amal Nala diversion seems unfeasible.

Further, it is significant to note here that nala diversion enhances mineral conservation in at least one mine (i.e., Naokari mine), thereby raising the overall mine productivity. This implementation can remove the production planning and scheduling difficulties for the remaining life of NLM mine. Another significant point that needs observation and evaluation here is about the implementation of the alternate mentioned the first and second points. Both of these alternates are *capital intensive*; hence, they are not the best solutions. Major or very large changes in the natural drainage course of the nala has negative impacts; therefore, total diversion is not desirable. On account of high expenditure and severe disturbances, the mine management may remain reluctant to apply such solutions. If the nala flow is altered partially (only in

S. No.	Parameter	BOP Nala	Amal Nala (Nala Vagu)
1.	Average dimension (W x D)	W - 15 to 30 m D - 20–60 m	W - 15–30 m D - 1–5 m
2.	Flow rate	Lean Period: 60–70 GPS Peak Period: 1800–1900 GPS	Lean Period: 50–60 GPS Peak Period: 1800–1900 GPS
3.	Drainage density	--	5.36 Sq Km.
4.	Other water sources (springs enroute)	Not present	Present, 03 prominent with flow rate of 1.2 m³ / second
5.	Highest flood level (HFL)	Not known	304.5 m
6.	RL of nala in lease area	200–210 m RL	300–308m RL

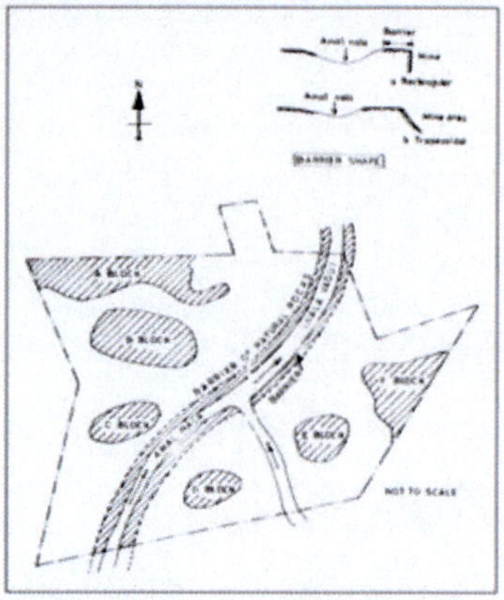

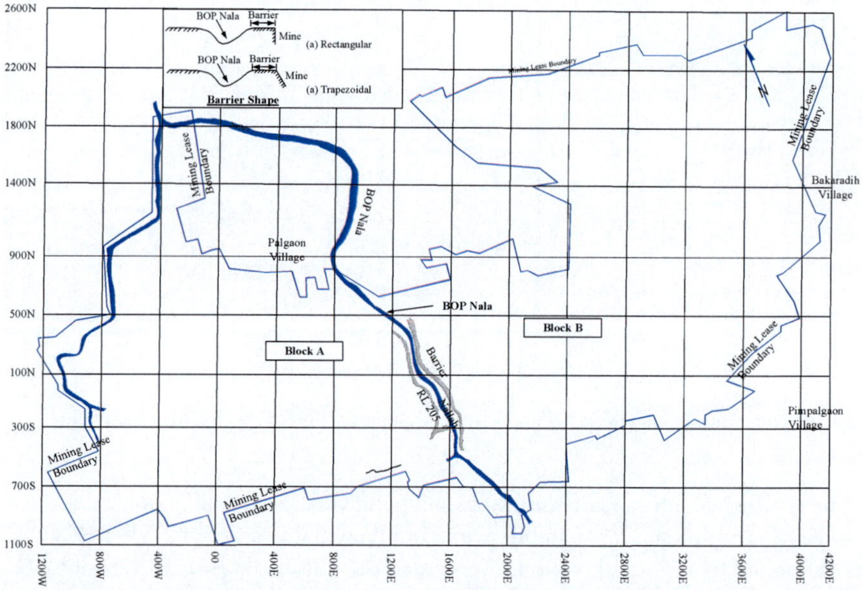

(b) Water channels (nala) in NLM and MLM mine areas and their salient particulars

Fig. 4.19 (continued)

the lease area), the effect on the natural drainage pattern of the nala or surrounding area seems insignificant because of two important facts:

1. The change in the nala course is for a partial length (route) only and not the full length. The confluence point of the nala with the Penganga River remains unaltered.
2. Impact of mining on the BOP Nala is localized and insignificant [15] because this drainage channel is seasonal and water flow is less.

Elevation levels [reduced level (RL)] of these surface water channels on the upstream and downstream sides are such that the water flow outside of the mine pit can be easily maintained with partial diversion. To implement this idea into practice, necessary statuary compliance must be adhered (i.e., prevention of water pollution in the nala). It is assessed that adoption of this approach will provide best engineering solution and also facilitates production planning and scheduling.

In this way, this case study allows us to learn that both limestone mine areas possess similar characteristic features.

(a) Intersecting nala (*BOP Nala* and *Amal Nala*) in both the leases
(b) Nearly same geological setting
(c) Same purpose of mineral extraction (i.e., cement manufacturing)
(d) Same mineral and method of extraction (i.e., opencast semi-mechanized mining).

In order to understand the mining conditions of both NLM and MLM mines in a better way, one should not lose track of *water management* and the *quality aspect of water* (water pollution). Without keeping these issues in mind, the assessment is not complete. In the proceeding paragraphs and in Annexure C, both of these aspects are explained. Annexure C is made separate for an easy and understandable description of the *hydrogeochemical analysis* of water.

With respect to the *water quality* (pollution-related aspects only) and *water management*, the following details are provided.

(a) The study of the drainage map of the area (see Fig. C3 in Annexure C) indicates that both of these mining lease areas have *dendritic pattern* of drainage.
(b) To make the mining operation environmentally friendly, it is necessary that care be taken at the mine level to contain and control water pollution.
(c) Annexure C presents water quality (WQ) evaluation and analysis in detail. For the ensuing case study, WQ analysis has been carried out using a *piper trilinear diagram*; *Gibbs variation diagrams*, and *U.S. salinity diagram* (Wilcox plot). With the help of parameters [namely *Sodium Absorption Ratio (SAR), Residual Sodium Carbon (RSC), Residual Sodium Bicarbonate (RSBC), Corrosivity ratio, Soluble Sodium Percentage (SSP), Kelley's Ratio, and permeability index,* shown in Tables AIII/3 and AIII/5 parts (a) and (b) of Annexure C], the suitability of water for different uses has been assessed.

Key highlights derived from the WQ analysis are described here according to water management in general.

Water Pollution

(i) The water quality assessment shows relatively high values of TDS, TH, SO_4^{2-}, Mg^{2+}, and F^- in some samples, making the water quality unsafe for drinking purposes. However, this water may be used for domestic purposes after treatment and disinfection. Based on SAR, RSC, Mg, and % Na, it has been concluded that most of the water in and around the studied mining areas can be used for irrigation without any hazards. It is also concluded that the prevention of fluorosis by *defluoridation* and *artificial recharge* are the most beneficial water management methods for these limestone areas.

(ii) The groundwater chemistry is mainly controlled by the weathering of minerals. Water-to-rock interaction, including the dissolution of carbonate minerals and silicate weathering, was the major hydrogeochemical process that affects the groundwater hardness. These results have been observed in the NLM and MLM mine areas as well.

(iii) Different geological domains (Penganga shale, limestone, sandstone, and Deccan basalt) where the mine area aquifer exists have varying rock porosity, which influences the groundwater movement within the mining area. In the NLM and MLM mine areas, groundwater quantity and groundwater movement are impacted due to the rock porosity present.

(iv) Water in and around the mining areas is alkaline in nature. The pH of groundwater samples from a dug well in the post-monsoon season indicates low-to-moderately alkaline nature of the water. The higher conductivity values (EC maximum 4280 μS/cm) in the groundwater of the study area are due to the high dissolved mineral percentage, indicating that the ion exchange and solubilization processes have taken place within the aquifer.

Water Management

This case study of two identical mines allows us to understand that best mining practice (BMP) and provides the best engineering solutions. The BMP here implies the continual improvement of existing methods [16] and should be adopted. Water is an essential and precious component for the NLM and MLM lease areas. Its uses should be optimized—whether for drinking, domestic, industrial, agriculture, or any other miscellaneous purposes. Related to water management and mining, the following points need careful attention for a smooth mining operation.

- Better mine design, improved planning, and optimum water use are possible, practical, and implementable.
- Good production planning and scheduling coupled with mineral conservation can be achieved without any problem by partial diversion of the BOP Nala.
- Proactive measures to make the water management effective in the mine areas are easy and needed.

In summary, the engineering solution of nala diversion for better mine planning, better water management, and pollution abatement is feasible at the studied mines. At the NLM and MLM mine areas, solving the water problem for both limestone production and mine productivity can be enhanced easily and conveniently. Taking cognizance of this case record for other mines elsewhere, we can benefit from this analysis. However, when plans for such nala diversion are executed, an in-depth evaluation remains necessary to find the best feasible solution.

4.3 Operation of Limestone Mines

From the case studies discussed, the following points are observed:

- In India, limestone occurs predominately near the surface; therefore, it is mined mostly by quarrying. The stripping ratio of a limestone deposit (limestone cover and overburden cover) is less.
- Mining operations at limestone mines are governed by several factors (Table 4.6). The physical geography and topography (plain deposit vs. hill deposit; coastal deposit vs. inland deposit) play significant roles in the design and development of limestone deposits.
- Scientific planning and mine designing is an essential constituent that serves the best purpose for systematic production and smooth operation for a complete mine life. The planning and designing of a surface limestone mine may look simple, yet it involves a large amount of exploration and various other data from geotechnical, environmental, economical, and mine planning angles. The grade control, mine scheduling, production planning, and optimization of core mining unit operations (i.e., drilling, blasting, loading, crushing or grinding, and hauling or transport optimization) are needed for successful production and mining of limestone.
- Deposit shape, deposit size (dimension), mineral thickness, orientation, and configuration are important during exploitation.
- The breakeven point at which the mining operation becomes profitable is a key component in limestone mining operation.
- Normally, in limestone mines, the production and consumption point (cement plant) are located close to one another, reducing the ROM transportation length. Since limestone is a mineral of low value, a smaller fleet is desirable.

- Conventional and mechanized systems of mining (continuous and non-continuous) have different planning and development milestones and schedules.
- Mine safety at the operational and every stage of limestone mines is extremely essential for mineral extraction. Shortcuts are not desirable and preferably are avoided for better safety at the workplace.
- All limestone mining operations should emphasize periodical skill upgrades. The use of software and modern techniques for information and communication (e.g., remote sensing, GIS, artificial intelligence, and 3D deposit modelling) should form an integral part of limestone mining.
- Basics of exploration and exploitation to deal with the complexities and variations of each deposit need to be applied for opencast limestone mines.
- Public concerns about environmental issues should be kept in view for smooth operation of mines.

Indeed, for totality, existing practices of limestone production from an open quarry should be such that the impact of mining on environment should remain at a minimum.

Table 4.6 Factors governing selection of surface mining methods for limestone

Limestone deposit conditions				
S. No	Characteristics	Open pit	Quarrying	Strip mining
1.	Ore strength	Any	Any	Any
2.	Rock strength	Any	Any	Any
3.	Deposit shape	Any (preferably tabular)	Any (preferably massive or thick bedded)	Tabular bedded
4.	Dip of deposit	Any (preferably low dip)	Any if thick	Any(preferably low dip)
5.	Size of deposit	Large thick	Large thick	Large moderate thickness
6.	Ore grade	Low	High (assay not critical)	Low
7.	Ore uniformity	Uniform (or sort of blend)	Uniform	Fairly uniform
8.	Depth	Shallow to moderate	Shallow to moderate	Shallow
Advantages and disadvantages				
9.	Mining cost	10% highest	100%	10%
10	Production rate	Large scale	Small scale	Medium scale
11.	Productivity	High	Very low	Medium
12.	Capital investment	Large	Small	Large
13.	Development rate	Rapid	Moderate	Moderate
14.	Depth capacity	Limited	Limited	Limited
15.	Selectivity	Low	High	Low
16.	Recovery	High	High	High
17.	Dilution	Moderate	Low	Low
18.	Flexibility	Moderate	Low	Moderate
19.	Stability of faces/openings	High	Highest	High
20.	Environment risk	High	Moderate	Very high
21.	Waste disposal	Extensive	Moderate	Minor
22.	Health and safety	Good	Good	Good

(continued)

4.3.1 Integrated Long-Term and Short-Term Planning of Limestone Mines

Lower production cost and the implementation of environmental management measures are achievable through integrated long-term and short-term planning of a mine. An integrated planning process, working in a hierarchy from long term to short

Table 4.6 (continued)

Limestone deposit conditions				
S. No	Characteristics	Open pit	Quarrying	Strip mining
23.	Others	Economical due to low breakage cost; rainfall and weather problems; best for large scale	Waste intensive; labour intensive; high-breakage cost	Easier waste and ore handling due to gravity; transportation and haulage problems due to hilly topography; less breakage cost; best for medium scale

Source Modified from Hartman [17]

term, is essential. In the limestone excavation process, developing mine plans based on geological and geotechnical information, generating mine designs, scheduling mining and limestone processing operations (crushing/sizing/mixing, etc.), and forecasting economic aspects (cash flows, market trends, and profitability) are the part of long-term planning to achieve a complete target of production. Working out the mine life, formally preparing blueprints for execution on a broader perspective, and getting this approved are also needed for the long-term plans for the mine. Thus, long-term planning clearly puts the focus on identifying and delivering the corporate goals. These plans set the overall strategic direction and are needed for the company that owns the mine.

The short-term plans provide more detail and precise data for real-time execution of the workforce (i.e., implementation). All short-term plans have interwoven aims in consonance with the goals of long-term plans and are at the front end of the achievement of long-term plans. Fixing or setting monthly and annual targets, value generation, and time-bound implementation of a desired set of objectives, following the underlying planning process, are important for the focus of short-term plans. The main differences between short- and long-term planning are thus the timeframe and the level of details and veracity. A planned mine may face technical problems from time to time, but it does not continually lurch into deep crisis. Most operational problems can be foreseen, avoided, and solved with good integrated long- and short-term planning (Fig. 4.20).

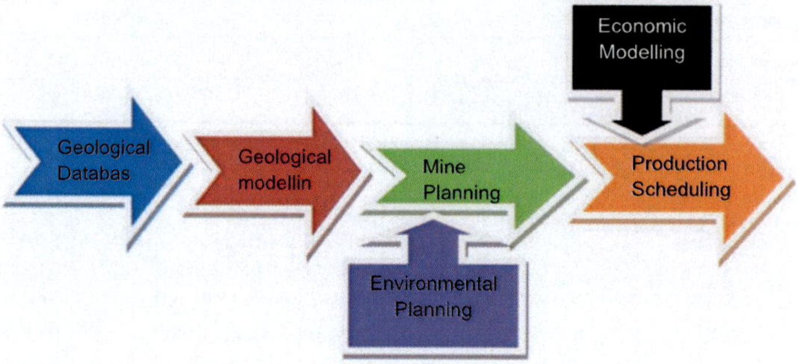

Fig. 4.20 Mine planning workflow—an integrated approach

4.4 Environmental Issues and Management in Limestone Mining

Environmental issues and their management in mining operation is one large area that encompasses several components. The major environmental issues in mining operation include (a) land degradation, (b) water pollution, (c) air pollution, (d) noise pollution and vibration problems, and (e) health and safety of workers [18]. A brief overview of each of these issues is given next.

Land Degradation

Land degradation is the most serious issue in mining and can be attributed to excavation of large scale, land subsidence, loss of vegetation and deforestation, and disposal of solid waste generated.

Water Pollution

Discharge of acid mine drainage and toxic wastes containing heavy metal and radioactive waste and the deposition of sediment load due to runoff from mining sites can be highly detrimental to the quality and health of the receiving water bodies in the vicinity of the mine. Further, underground mining can result in lowering the water table, and mine water can infiltrate into the underlying aquifer and pollute groundwater.

Air Pollution

Air pollution is also a serious problem in mining. Dust is generated from various mining activities via excavation, loading, unloading, and transportation of minerals and ore by vehicles, whereas gaseous pollutants (i.e., sulphur dioxide, oxides of nitrogen, and carbon monoxide) are generated from dumpers and other transporting vehicles. In limestone mining, *fugitive dust* is also one problem as a source of air pollution.

Noise Pollution and Vibration

In limestone mining, there are three major categories of noise sources: fixed plant installations, mobile plant units, and transport movements (external). Heavy earth-moving machinery (HEMM), drills, dumpers, and material handling are the prominent noise sources in limestone mines.

Similarly, vibrations due to HEMM, equipment, and mine blasting are prominent and significant in a limestone mine. These issues need proper management measures for better environmental health and safety (EHS).

Noise and vibration pollution is damaging because it interferes with speech communication and can cause distraction, nervous irritability, hearing damage, and physical strain.

Occupational Health and Safety

The main influential issues concerning the occupational health and safety of miners include their poor socioeconomic status, living and working conditions, and lack of general good health. The living conditions in mining areas generally remain far from satisfactory. It is known that mining is an accident-prone activity. Mine workers are always exposed to probable risk of accidents that are caused either by human error or failure and defects in machinery. Note that both health and safety are important for this industry.

Environmental Impact

To evaluate the magnitude and significance of the potential environmental impacts from the mining process, an Environmental Impact Assessment (EIA) is necessary. EIA identifies, predicts, and evaluates likely environmental impacts. Based on that, an Environmental Management Plan (EMP) is prepared for minimizing/mitigating the impacts. EIA/EMP studies for limestone mines encompass baseline data collection for all the important components of the surrounding environment, including air (ambient and workplace); water; land; noise and ground vibration; solid and hazardous waste; and biological environment (flora and fauna) as well as socioeconomic aspects through field visits and monitoring. The field and lab data is analysed to assess the existing quality of environment for different stages of the mine (prior to commencement of mining, during mining, and at the post-mining phase). EIA/EMP studies also predict future scenarios due to proposed project activity through the use of various modelling tools and delineating necessary remedial and mitigating measures. The assessments will also help develop a post-project environmental monitoring program in order to maintain the quality within the regulatory standards and limits prescribed for each component of environment. A *Risk Assessment and Disaster Management Plan* is also an integral part of any EIA/EMP. In India, any developmental activity, including mining, has to comply with the rules and regulations contained in the EIA Notification of 2006 as well as the Environmental Protection Act of 1986 and notify the Ministry of Environment, Forest and Climate Change (MoEFCC) prior to commencement and also while in operation.

Field studies (along with the water quality standards, emission standards (air quality), and ground vibration/noise norms) help the mine management in framing environmental programmes for their mines and general organization. The mine management makes future plans based on the Comprehensive Industry Document (COINDS), which is prepared as per the environment impacts caused at site. For protection of the environment, the EMP framed for a particular mine is generally obtained to achieve environmental excellence. Several devoted indigenous organizations and experts from the mining industry provide valuable inputs according to the needs of mine.

In addition to the environment components, other studies, such as a socioeconomic environment for quality-of-life assessment, life-cycle assessment (LCA) studies, and carbon footprint assessment, are equally significant today. Each of these areas are very broad, so their detailed descriptions are not included in this book.

From these descriptions, it is evident that to assess, evaluate, and manage the environmental impacts of mining limestone, consciousness and sincere efforts are needed. The uninterrupted operation of a limestone mine can be ensured only with a proper balance between the mine development and the surrounding environment.

4.5 Parivesh: A Single-Window System for Limestone Mining Clearances in India

In order to bring more transparency and accountability to the forest, environment, and wildlife clearance processes for any developmental project of limestone mining, the Ministry of Environment, Forests, and Climate Change (MoEFCC) of the Government of India has developed a portal called the Pro Active Responsive facilitation by Interactive and Virtuous Environmental Single-window Hub (PARIVESH). PARIVESH is a web-based, role-based, G2C, and G2G workflow application that was developed for online submission and monitoring of the proposals from the user agencies for seeking environment, forest, and wildlife clearances. It automates the entire tracking of proposals, which includes online submissions, editing or updating the details, and displaying the status at each stage of the workflow.

The term G2C refers to *government to citizen,* and G2G refers to *government to government* and are used in business management for information from the industry or citizen in any life situation or a transfer of an official document to the person concerned. These abbreviations are generally used for the simplified communication in the electronic form of an office website or public service catalogue. The PARIVESH system is based on the web architecture. It uses IIS as an application server, Internet as a framework, and SQL as a database server (https://parivesh.nic.in).

References

1. CMRI (2007) Geo–hydrological studies for Kovaya limestone mine of M/s, UltraTech Cement, Gujarat Cement Works (GCW), District—Amreli (Gujarat), Technical Report No.: GC/MT/N/18/2005–06, (Unpublished), Central Mining Research Institute (CMRI), Dhanbad, Mar, p 103
2. CSIR-CIMFR (2017) Salinity ingress study with special reference to hydro-geological regime in and around Narmada Cement—Jafarabad Works (NCJW), UltraTech Cement Limited, Village—Babarkot, Taluka—Jafarabad, District—Amreli (Gujarat), Technical Report No. CNP/N/ 4445/2016–17, (Unpublished), CSIR-Central Institute of Mining and Fuel Research (CSIR-CIMFR), Aug, p 117
3. CSIR-CIMFR (2012) Assessment of technical feasibility of mechanical mining for Alsindi Limestone Deposit of Himalaya, District Mandi, Himachal Pradesh, Technical Report No. CNP/N/2781/2010–11, (Unpublished), CSIR-Central Institute of Mining and Fuel Research (CSIR-CIMFR), Apr, p 70
4. Allen SK, Allen JM, Lucas S (1996) Concentration of contaminants in surface water samples collected in west-central impacted by acid mine drainage. Environ Geol 27:34–37
5. Bhuiyan MA, Islam MA, Dampare SB, Parvez L, Suzuki S (2010) Evaluation of hazardous metal pollution in irrigation and drinking water systems in the vicinity of a coal mine area of North-Western Bangladesh. J Hazard Mater 179(1):1065–1077
6. Choubey VD (1991) Hydrological and environmental impact of coal mining, Jharia Coalfield, India. Environ Geol 17:185–194
7. Khan R, Israili SH, Ahmad H, Mohan A (2005) Heavy metal pollution assessment in surface water bodies and its suitability for irrigation around the Neyevli Lignite Mines and associated industrial complex, Tamil Nadu, India. Mine Water Environ 24:155–161
8. Soni AK (2007) Evaluation of hydro-geological parameters associated with limestone mining: a case study from Chandrapur India. In: Mine water and the environment, vol 26, IMWA, Springer Verlag, pp 110–118
9. Chan-Hojeone (2001) Mineral-water Interaction and hydro-geochemistry in the Samkwang Mine Area, Korea. Geochem J 35:1–12
10. Gupta MK, Singh V, Rajwanshi P, Agarwal M, Rai K, Srivastava S, Shrivastav R, Dass S (1999) Groundwater quality assessment of Tehsil Kheragarh, Agra (India) with special reference to fluoride. Environ Monit Assess 59:275–285
11. Rawat NS, Viswanathan S (1990) Assessment of water quality deterioration in some coal mines of North Eastern Coalfields and Jharia Coalfields. J Mines Metals Fuels 15–21
12. Singh AK, Mahato MK, Neogi B, Singh KK (2010) Quality assessment of mine water in the Raniganj coalfield area, India. In: Mine water and the environment, vol 29, Springer Verlag, pp 248–262
13. Taranekar PS (1993) Study of environmental implications of water quality with special reference to geology and mining activities in parts of Nagpur and Bhandara Districts of Maharashtra, (Unpublished Ph.D. thesis), RTMNU, Nagpur University, p 175
14. Tiwary RK (2001) Environmental impact of coal mining on water regime and its management. Water Air Soil Pollut 132(1–2):185–199
15. CMRI (2006) Geo-hydrological studies for Naokari limestone mine of Awarpur Cement Works, UltraTech Cements, District-Chandrapur, Maharashtra, Technical Report No. GC/MT/N/14/2004–2005, (Unpublished), Central Institute of Mining Research (CMRI), Feb, p 117
16. Soni AK (2017a) Mining in the Himalayas: an integrated strategy. CRC Press/Taylor & Francis, p 225
17. Hartman HL (1987) Introductory mining engineering. Wiley-Interscience, New York, p 570
18. Umarthay RM (2002) Textbook of mining geology. Dattsons, pp 94–121
19. CMRI (2000) Feasibility of Manikgarh cement limestone mine workings below ground level (303 MRL) and its impact on water regime of the area, Technical Report No.: GC/MT/R/1/99–2000, (Unpublished), Central Mining Research Institute (CMRI), July, p 64

20. CSIR-CIMFR (2009) Upper catchment area treatment plan for Nongtrai Limestone Mine of M/s Lafarge Umiam Mining Private Limited (LUMPL), Shillong, Meghalaya, Technical report No. GC/MT/N/16/2007–08, (Unpublished), July, p 134
21. CSIR-CIMFR (2013) Geo-hydrological study for mining at deeper levels in limestone mine of M/s Trinetra Cement (India Cement), District—Banswara Rajasthan, India, Technical Report No. CNP/N/3259/2012–13, (Unpublished), CSIR-Central Institute of Mining and Fuel Research (CSIR-CIMFR), Aug, p 72
22. CMRI (2006) Geo-hydrological studies for naokari limestone mine of Awarpur Cement Works, UltraTech Cements, District-Chandrapur, Maharashtra, Technical Report No. GC/MT/N/14/2004-2005, (Unpublished), Central Mining Research Institute (CMRI), February, p 117

Chapter 5
Environment-Oriented Development of Limestone Mineral-Bearing Areas

Eco-friendly mining is a recognized buzzword in the mining business. An ever-growing awareness about the *mining environment* has drawn the attention of mining businesses and forced them to look at the environmental performance of mines more critically. In India, the demand for limestone as a major mineral is consistently rising; therefore, prospects of environmental management with respect to a mine have become more and more challenging. The cost associated with environmental management has remained a subject of discussions for the environment-oriented development of mines—even with regulations in place under legal obligations. Since economics are anchored in terms of improving the productivity of individual mines, it is always desirable that shortcuts be avoided. In the present—as well as for a vision beyond 2030 scenario—India's limestone mineral wealth can be easily and efficiently exploited for sustainable economic growth provided due care is taken for environment protection.

For limestone-bearing areas, environment-oriented development commences from geological exploration and ends at the 'green' closure of mines. At every stage, healthy and clean surroundings matter. This chapter has a discrete focus on the integrated management approach and economical solutions in limestone mining. A proactive approach in managing the environmental aspects of mining makes good business sense for the concerned organization. Various aspects of limestone mining environment (e.g. conservation, value addition, and societal consent) have been discussed in order to highlight the importance of environment management in a holistic way.

5.1 Sustainable Development (SD)

Sustainable development and *environment* are well described but less religiously implemented words. These days one more word is added to these two words: *climate change* (Cc). The *climate change* is most vulnerable, widespread, and

A. K. Soni and P. Nema, *Limestone Mining in India*, Materials Horizons: From Nature to Nanomaterials, https://doi.org/10.1007/978-981-16-3560-1_5

causing unprecedented impact on environment and society. Climate change is most widespread and causes unprecedented impact on the environment and society. Impacts from the Indian mining industry— with particular reference to limestone mines of India—place additional burdens on climate change. Here is a basic definition of sustainable development.

> *We have to preserve our earth's environment, which includes mine environment as well, genuinely for future generation without damaging it. At every step of the value chain in limestone mining, environment has to be such that it facilitates well-being for humans and brings ease of working in all mining activities (i.e. starting from exploration to exploitation to ore processing and finally ore dispatch to its consumption destination). Technology imperatives would be a great saviour in maintaining a clean and green limestone mine, and use of NexGen technologies will be a tool to rely on for promoting sustainable development and to be implemented through responsible mining.*

Now and in the future, *environment* and *development* are and will remain inseparable. Industry cannot survive without protecting the environment while promoting development. The fact that the mining of minerals causes a negative impact on the environment is a reality in the truest sense; therefore, society and people have to strive hard and face the numerous challenges that lie ahead.

Sustainability is an act of maintaining development that should remain at the core of any mining operation. This can be achieved by applying ecologically sound mining practices with the technologies involved in the process. Prevention and control of emissions and effluent and wastes through using the best mining practices are achievable provided mining operations comply with applicable legal and other statutory requirements as laid down by various concerned government bodies (MOEFCC, CPCB/SPCB, IBM, CGWB, etc.). Additionally, safety norms laid down by the Directorate General of Mines Safety (DGMS) for internal safety rules and standards of companies and their mines are equally important for carrying out sustainable limestone mining. In addition, *positive thrust on quality* and *societal development* are equally necessary for any company. Sustainable development is accomplished only when the improvement in the quality of life of people in general and the upliftment of society in and around the mines in particular is clearly observed.

5.1.1 Sustainable Development Goals (SDG)

For mining to be sustainable, self-regulation(s) and individual action(s) are more important than statutory compliance. To achieve the environment-oriented development of limestone mineral-bearing areas, goals such as scientific planning; cost optimization; mine productivity; workplace safety; low-grade ore beneficiation (through identification of new exploration techniques, including economical processing); local cost-effective solutions; short- and long-term environmental management of mining operation and related industrial practices; pit slope management; and environmentally friendly transportation systems are needed.

Fig. 5.1 Sustainable development goals (SDG)

Since the dawn of time, there has been a constant conflict between humans and their environment. Mining teams often face the dilemma of priorities—*production first* or the *environment first*. Sustainable development examines and provides answers to this dilemma from both business and environmental perspectives. Growing awareness in society is spearheading the accelerated thrust on environment protection today.

Environmental laws give citizens their basic constitutional rights to live in a clean environment. Urgent action to slow down or even halt the caustic effect of a mine's environment on climate changes should be an integral part of sustainable development goals (Fig. 5.1) fixed for society.

5.1.2 Green Credit Scheme (GCS)

The Forest Advisory Committee (FAC) of the Ministry of Environment, Forest and Climate Change (MoEFCC) of India is an apex body that examines and considers issues relating to the diversion of forest land for non-foresting uses, such as mining, industrial projects, and township expansion. Also issues granting forest clearances have been approved as a *Green Credit Scheme* (GCS) on 09 January 2020. The FAC approval allows forests to be traded as a commodity. This is not the first time that such a scheme has been mooted. In 2015, a Green Credit Scheme for degraded forest land with public–private participation was recommended. However, it was not approved by the MoEFCC.

As per the current system, the industry should find appropriate non-forest land equal to that which would be used for industrial purposes (diversion of forest land for industrial uses, including mining, which is generally done in unpopulated forest areas). Industry needs to compensate for such diverted land by finding non-forest land and developing an equal amount of forest in that land. However, industries have often complained that they find it hard to acquire alternate non-forest land, which has to be contiguous to an existing forest. In brief, the positive practice is for the loss of forest to be replaced through *compensatory reforestation* at an appropriate place or location by planting suitable local trees and plant species.

The current GCS system allows the forest department, either at the state level or central level, to outsource afforestation to non-government agencies (NGOs) for GCS implementation. The GCS allows individuals and different agencies, like private companies and village forest communities, to ensure afforestation and its practical implementation. The principal job of a forest department task force would be first to identify lands for forest cultivation and ensure that plantations are grown. GCS can be implemented successfully through public–private partnerships. After three years, these attended identified lands would be considered compensatory forest lands after meeting the forest department's criteria.

An industry that needs forest land could approach to the state or central ministry, as the case may be and pay them for it. This would then be transferred to the forest department and also recorded as *forest land*. The economic equivalent of the forest land is called the net present value (NPV), which should be paid to the state forest department. According to statistics, in India, nearly ₹50,000 crores have been collected by the MOEFCC in the past decades, but the funds remained unspent because states did not regrow the forests. It was only after the Supreme Court's intervention that about ₹47,000 crore was disbursed to those states in August of 2019.

GCS is helpful to meet international commitments, such as sustainable development goals (SDG). Through plantation use and compensatory afforestation–either by individual or through community participation—social and ecological changes are achievable.

5.2 Integrated Management Approach

Management encompasses academic rigour and a practical approach for mining, as well as for the mining environment. It explores and analyses local issues within a global perspective. Thus, it is commonly referred to as an *integrated environmental management* (IEM) approach. Solutions provided by this approach are commensurate with global, theoretical, and methodological requirements. The broad matrix of an integrated management approach in selected and limited wording is as given in Table 5.1, which depicts the combination of engineering and management.

Table 5.1 Matrix of integrated management approach

Management	Engineering
Technical problem Social problem	Energy Mechanical, electrical, mining, and social science engineering
Legislation and compliance	Mining of mineral (limestone) Waste
Policy	Corporate versus unit Environmental engineering
Costs and taxes	Production engineering Cost and accounts; income and taxation
Standards	Materials Industry-specific standards
Pollution	Air, water, land, and noise Environmental auditing
Impacts	Products, mining, and environment
Human resource development	Education
Information dissemination	Computers and information
Organizational commitments and reviews	Different related disciplines of engineering

The integrated approach to the environmental management of limestone mines encompasses air, water, and land pollution arising out of mining activities. This in turn has links with the regional, national, and global effects of pollution leading to acid rain, ozone depletion, global warming (on a wider scale), eutrophication of water bodies, and so on. To conserve resources, an increase in efficiencies and the reduction in generated mining waste are some key principles of IEM (Table 5.2). This integrates environmental management with environmental engineering and covers various aspects of mining with respect to energy, pollution (air, water, and noise), the use of materials, the production of waste (effluents), etc. Another aim of IEM concerns problems and solutions; legislation and circular economy; policy and standards; impacts and mitigation; and economics and profit for the prediction and assessment of pre and post-mining scenarios. These are influenced significantly with the commencement of mining operations. Thus, IEM is a novel approach to developing an optimized decision-making sequence for the best possible mineral production, energy utilization, and conservation of resources. Based on the list of management decisions, the cost-effectiveness of the limestone mining operation is formed and implemented.

Therefore, it is clear that a systematic application of an integrated approach in mine management—as well as environmental management of mining operations—allows all necessary alterations and modification (changes) to technical activities being resourced financially in a project. An illustrative case study may be referred, which can demonstrate the efficiency, effectiveness, and financial value of an adopted approach [1].

Table 5.2 Key principles in environmental management of limestone mines

Sustainable	Environment and development should go together with an eye on protecting present and fostering future generations in terms of resource conservation and protection. Policies should contribute to the goal of sustainable development
Contained (protection at source either *point source* or *line source*)	Environmental damage should be contained or rectified at the source as a priority
Precautionary	Where outcomes are uncertain, particularly if they are likely to be irreversible, there should be a presumption in favour of a precautionary or cautious approach
Integrated	Decisions concerning environmental impacts should have regard solutions (options for all media) and be taken in an integrated and holistic way
Effective compliance	Decisions should be implemented at the appropriate level (i.e. grassroots) or at the local scale, be reasonable in the circumstances, and recognize that the diversity of situations applying in different regions will lead to different practices being applied. Effective compliance followed by necessary perusal or follow-up is important—even if environmental protection is the cause of general public concern

5.2.1 Social License of People to Operate

Today opposition from the local population to large developmental projects, including mines, becomes *de rigueur.* On one side, mining company faces the execution difficulty because mining is a tough job while on social front, local people object to mining and create many hurdles.In order to address and resolve this public issue, it becomes imperative that the local population is consulted and their consent obtained before any project begins. In India, a clause requiring a *public hearing* for the new mines or for expansion of production of existing mines was added in the system [2]. The purpose of a public hearing is to obtain consent from the local people in order to operate a mine smoothly. Based on past experiences of the Indian mining industry, the very purpose and constructive approach of a public hearing is an understood method of obtaining positive feedback on environment protection and social benefits. Unfortunately, these hearings have been altogether left behind and turned out to be an employment blackmailer in over-populated developing countries like India, where jobs are in short supply.

Mining in India needs a humane approach because of several social factors (i.e. literacy, unemployment, and poverty) at grassroots levels. While the mining minerals are needed to fuel the economic engine of country (GDP growth), projects that affect tribal communities and forest resources should not be ignored, especially when we say that it is an integrated development. Thus, a sustainable mining policy stated in the National Mineral Policy of India will bring employment and environmental protection together. More importantly, the state and mining companies must follow the rules (in letter and spirit) when obtaining consent from the people concerned and use their resources, such as the district mineral fund (DMF), for the betterment of the region and community living in that area.

In the future, it will not be easy for any limestone mine operator—small or big—to run a mine without a licence from society that requires the support and consent of local people. The mine managers will be required to adopt an integrated approach that will improve production and ensure socioeconomic and other relevant benefits to the people at large in the project region.

5.3 Environment-Oriented Development: Alternative Energy Sources and Waste

In order to provide environment-oriented development, raw materials (limestone, coal, etc.) have to be consumed judiciously. In Sect. 2.6, the *circular economy* concept was explained. In a circular economy, the use of raw materials is optimized, and the intelligent reuse of any waste products created during mining, manufacturing, and processing is promoted. This section discusses the issue of alternative energy sources and waste because coal is a mineral material input in cement making. Why coal? The answer is that coal is also produced from mines (like limestone), and it is a direct energy source. Therefore, coal is part and parcel of the cement-making process and is the main source for heating limestone in kilns. In limestone mining, overburden handling and waste management are not important issues because normally there is no significant overburden or soil cover encountered over limestone deposits in India. Further, in order to reduce the consumption of coal, municipal solid waste (MSW) available from the surrounding areas of a mine could be used as an alternative fuel for heating the limestone kilns. Another option could be using the plastic waste (Box 5.1). These solutions take into consideration the related air and water pollution issues and necessary safeguards. What should be the percentage of alternative fuels used? Replacing coal and fossil fuels has to be worked out through proper scientific studies covering economic and environmental aspects.

5.3.1 Alternative Energy Sources

The term *thermal substitution rate* (TSR) is used in the context of alternative energy sources that replaces coal as a conventional resource. This term refers to the percentage of sustainable alternative fuels used to replace fossil fuels. Currently, the average TSR in the cement industry has moved up to 3% from less than 1% about a decade ago. The limestone mining industry is now working towards reaching a TSR of 25% by 2025 and 30% by 2030.

Box 5.1: Municipal Solid Waste (MSW)

Municipal solid waste (MSW) includes material that is combustible in nature but not recyclable, such as soiled paper, cloth, contaminated plastics, packaging materials, leather, rubber, tyres, polystyrene (thermo coal), and wood.

According to the United Nations (UN) forecast, the urban population in India will grow at a rate faster than China, which is presently the highest populated country of the world. As per the current estimates of the Central Pollution Control Board (CPCB), the amount of municipal waste generated in India stands at an alarmingly high rate (144,165 metric tons/day). Of this, about 40–60% is comprised of organic waste, while the rest is inorganic. MSW is estimated to be growing at a rate of 1.33% per capita per annum and has become a cause for national concern. The corresponding greenhouse gas emissions are also very high and are expected to grow in the future in a *business as usual* (BAU) scenario—from 19 to 41 million t CO_2e by 2030.

Box 5.2: Plastic Waste

Millions and millions of tonnes of plastic waste are produced in India which is enough to fire rotary kilns of cement plants, resulting in the conservation of conventional fuels, such as coal. By doing so, up to 10% of energy savings can be achieved annually by 2025. According to the Cement Manufacturers Association (CMA) of India, Indian cement plants have the potential to consume 12 million tonne of plastic waste annually [5].

The enhancement of the TSR requires adequate preprocessing infrastructure for alternative fuels that will make it suitable for use in captive cement plants where the limestone mines are operative. One of the alternatives to make use of locally generated waste is to blend it with industrial waste. The mixed waste when used as process feed could yield a suitable alternative fuel that can produce required energy to run kiln. A fuel which has a consistent calorific value of around 3600 kcal/kg is needed for firing cement kilns, and blending of commercial and industrial waste in the ratio of 55:45 can meet this requirement.

The boxed descriptions highlight the importance as to how many waste products can be used as alternative energy sources (as substitute for conventional fuel like coal) in the cement industry which would achieve the dual targets of resource conservation and environment protection. It is now well established that the use and waste minimization of alternative fuels offer great opportunities for environment-oriented development. Thus, the conservation of minerals through substitution of alternative energy sources is an environment protection-oriented approach that needs to be given due consideration.

5.3.2 *Limestone Conservation*

According to the National Mineral Policy of 2019, the conservation of minerals shall be construed not in the restrictive sense of abstinence from consumption or preservation for use in the distant future but as a positive concept leading to the augmentation of reserve or as a resource base. There shall be an adequate and effective mechanism and framework for promoting *zero-waste mining*, as its ultimate goal is to conserve minerals by preventing wastage [3].

It is true that minerals conserved means minerals produced. Therefore, all companies engaged in limestone (or mineral) production should include this aspect in their policies for environmental benefits as well as for ever-increasing production rationale. Although limestone is a mineral of low value, there is still a need for conservation practices in mines for inclusive growth. In fact, the situation created by the mining activities at a mine or project site causes great impact on different conservation practices, which are linked to SDGs. In India, the Mineral Conservation and Development Rules (MCDR) of 2017 are in place, requiring the registration of mines under Rule 45. This promotes the conservation of minerals being mined statutorily. The mining community needs to give appropriate priority to mineral conservation due to the fact that *a mineral is a wasting asset*. If exploited all at once, it will take million of years for its formation again. Although limestone is a rock in abundance, its sedimentary origin and formation took several years. Hence, we must conserve it. Unfortunately, this aspect is given least priority in the day-to-day working of mines because of production pressure. Nevertheless, limestone conservation is indeed a desired aspect of limestone mining.

5.4 Extracting Value from Limestone Mining Operation: Some Innovative Concepts

To extract value from a limestone mining operation, the end use of the mine be ascertained first. Such end use of the mine is ascertained through the *closure plan* of the respective mine or mining enterprise. Without waiting for the mine to be closed fully, part of the mine area that is abandoned or from where mineral is exhausted is where the environmental restoration of the land area shall be done. This includes development of the area and putting it to various uses (e.g. for recreation—say a playground, for pit lake development, for rainwater harvesting, for groundwater recharge, for pisciculture (fish breeding), or for any other civic uses in demand for that area.

Limestone mining companies around the globe are looking for smarter solutions and with a deeper commitment towards strategies for reducing the *waste generated* during mining activities. *Reprocessing* is a viable solution which helps to minimize risk and deliver a valued return on investment. For example, limestone mining generates a considerable amount of dust. This dust can be utilized for making agglomerate or pellets that can be used as sweetener or raw material feed where a higher $CaCO_3$ product is required, or it also can be used as building material.

5.4.1 Mine Water Value

Limestone mining operations are most often limited to open-pit mining. All of these surface mine, operational, or mined out pits from where limestone is extracted contain huge quantities of stored water referred to as *mine water*. Such water storage occurs naturally in the open pit due to a puncture of the water table (during mining below the water table) or accumulated rainwater. This mine water can have useful purposes for which a number of novel and practical ideas are available, as listed here.

(a) Mine water can be used for improving coal quality by washing the coal that is used in a kiln for producing cement. In one cement plant that uses coal for firing a cement kiln, pre-washing the coal using mine water derives the benefit of reducing coal consumption and lowers ash generation at the source. The operational efficiency of the kiln can be enhanced considerably with less coal consumption and more energy use. Various captive mines of cement plants that are owned by private companies can get commercial benefits from this idea by making effective use of mine water within their industrial areas for a selected specific purpose.

(b) Mine water can be used in construction related to civil works in mines and plants.

(c) Mine water can be used for wetting roads at mines and cement factory campuses to suppress dust. This will save operational costs.

(d) Mine water can be used for plantation and greenbelt development, which is a statutory requirement of a mine area.

(e) Water is *saleable*. Both treated and non-treated mine water can be sold by the mining company for miscellaneous purposes to local users. Extracting value from mine water is found by using it in irrigation and agriculture. In an arid region of Rajasthan, India, this may be of great benefit.

(f) Mine water can be made potable after a *water quality analysis*. Thus, the portability of mine water is one area that also extracts value from mine water. A packaged drinking water plant near the mine is an example. One such successful example is found at Western Coalfields Ltd's Saonar mine in Maharashtra state of India [4].

(g) There are miscellaneous uses in mines and cement plants, colonies, gardening, etc., for mine water.

(h) In captive mines of nearly all companies and cement plants, *water charges* are paid to the government by the industries for the withdrawal of fresh groundwater for various uses (i.e. industrial operations, colony supply, etc.). The freshwater use can be reduced when mine water is used for non-potable purposes, providing a considerable savings.

(i) Mine water can also help in overcoming a *water crisis* (if such situation arises). Bore wells and water collected in the mine pits can meet the water requirements of the cement plant of a captive limestone mine located in scanty rainfall areas. Moreover, the barren land available to the company can be used for constructing *rainwater harvesting structures*. This initiative results in an increase of water availability for use in the lean period in areas like Andhra Pradesh Cement Works (APCW) in the district of Ananthapur, Andhra Pradesh, India (Fig. 5.2)

(j) If a mine is located in a coastal region near the sea and the sea water filters inland, making mine water salty (such as in the limestone mines of Gujarat Coastal area), the pit water can be converted into *salt pans* to make salt.

Since the availability of water in required quantities is always a question mark for any industry, effort should be made to conserve it and use it judiciously. In particular, for the mining industry, mine water should not be wasted. It could be properly used to save freshwater, thereby offering an opportunity for additional value. The value addition from the mine water can be easily and effectively implemented into practice using a corporate social responsibility (CSR) scheme where the mining company concerned (Chap. 3; Sect. 3.3). These practical approaches for the conservation of water are not merely value addition-oriented but are essential for the conservation of natural water resources.

5.4.2 Mineral Matters: Limestone and Cement Types

Limestone (as a mineral) is the main ingredient for making cement, and its quality in terms of carbonate percentage has great significance. Finely ground limestone is a highly sustainable material with a wide array of uses. State-of-art research has demonstrated that if the grinding of limestone is optimized, Pozzolana Portland cement (PPC) can be produced. This PPC will have a similar performance to that of Ordinary Portland cement (OPC), leading to both significant cost savings and carbon footprint savings, without compromising quality.

For value addition to limestone produced or to extract value from limestone mining operations, different types of cement have been developed. A short description of each of the most common types is given next. From this, readers can assess and understand how the limestone can be optimally and alternatively used as a resource.

1. **Novel Cement**: Some cement manufacturing companies, based on their R&D, have produced of *novel* cements, which eliminate the need for Portland clinker altogether. Novel cement products are competitive with the production cost and performance of *Portland cement*. Moreover, the entailing emission reduction

Fig. 5.2 Saving rainwater through rainwater harvesting efforts at APCW. *Source* [6]

is quite significant. However, it may be noted that these products have not yet achieved large-scale commercial use and are limited to niche applications only. However, good potential exists for such products, and they are expected to go a long way in the future.

2. **Geopolymer-based Cements**: This kind of cement has been a focus of research since the 1970s. This cement does not use calcium carbonate as a key ingredient. Geopolymer cements harden at room temperature and release only water. Compared to Portland cement, around 80–90% reduction in emissions is possible in its manufacture. Two companies based in the United Kingdom (Zeobond and Banah) are engaged in manufacturing this cement.

3. **Carbon-Cured Cement**: *Carbon-cured* cements are carbon negative products—those cements that absorb CO_2 rather than water as they harden. If this CO_2 absorption can be made higher than the CO_2 released during their production, these cements could potentially be used as a carbon sink. Several firms are developing this type of cement. The United States firm Solidia claims that its cement emits up to 70% less CO_2 than Portland cement, including this sequestering step. The firm is now in a partnership with major cement producer Lafarge Holcim. There are other firms that are using completely different materials to make this type of cement.
4. **Bacterial Cement**: North Carolina-based start-up firm Biomason, for example, uses bacteria to grow cement bricks that it claims have the same strength as that of traditional masonry and carry carbon-sequestering potential as an added advantage.
5. **Fly Ash–limestone Cement and Slag–limestone Cement**: In India, fly ash and blast furnace slag have been used as *additives* to limestone in cement making. This is method that allows limestone use to be optimized in the production of cement. Such practices are now common in many cement plants of India and are covered by the Indian cement manufacturing standard. Due to its abundance and ease of processing, these two materials have gained popularity.
6. **LC3 Cement**: LC3 is a cement type called *limestone calcined clay cement*. This cement type was developed about five years ago at the EFPL research institute in Switzerland. Basically, LC3 is a formula where the quantity of clinker is cut almost in half by adding calcinite clays and limestone—both of which are cheap and often abundant. LC3 cement also can be cooked at much lower temperatures (around 700–800 °C compared to the 1400–1500 °C needed for clinker manufacturing), thereby saving up to 30% off its carbon footprint compared with Portland cement. This cement type has already been commercialized, but it is not yet being adopted at a large scale due to a host of barriers, including inflexible industry standards and concerns over how new blends will hold up long term [7].

In conclusion, it is clear that newer tec*hnologies and recent advances* in material science would be quite promising to the additional value of limestone. Its large-scale use as a raw material for various products, as a construction material, or as industrial filler material will signify how limestone matters.

References

1. O'Callaghan PW (1996) Integrated environment management handbook. Wiley, England, p 386. ISBN 0-471-96342-9
2. Saxena NC, Singh G, Ghosh R (2002) Environmental management of mining operation. Scientific Publisher (India) Jodhpur, p 410. ISBN 81-7233-296-3
3. GOI (2019) National Mineral Policy (NMP) 2019, Ministry of Mines, Government of India (GOI), p 12

4. Soni AK (2019) Mining of minerals and groundwater in India. In: Gomo M (ed) Groundwater-resource characterization and management aspects. IntechOpen, pp 71–103. https://doi.org/10.5772/intechopen.73345
5. CMA (2019) Cement Manufacturers Association (CMA), Noida, India. https://www.cmaindia.org/, Official Website. Accessed on 11 Dec 2019
6. SDR (2019) Driving growth through SDGs, UltraTech Cement sustainability development report (SDR) 2017–18, Final web version, Feb 2019, p 62
7. Karen L, Scrivener (2014) Options for the future of cement. Indian Concr J., July-Special issue-Future Cement, pp 11–21

Chapter 6
Modern Technological Applications for Limestone Mining

The mineral industry in India is undergoing a technological upheaval that is usually referred to as *modern-day reforms* or *modernization*. In various mining organizations, this has been reflected in terms of some noted characteristic features, such as exchanging knowledge, upgrading existing practices, and allowing private participation in production, including the expansion of production and international collaborations. Considering this trend, the core of this chapter has been written.

The technological application is an exhaustive, broad, and topical subject that is focused here on limestone mining and its ancillary aspects. The *specifics of mining* or the uniqueness of situations (field conditions) that necessitate the selection of a combination of old and modern methods and equipment also will be discussed. Mine operators, including decision-makers, should observe grounded realities of mine sites and create interdisciplinary and tailor-made solutions for troubleshooting problems. There may be multiple benefits of such solutions instead of one.

6.1 Scientific Planning

Commercial production of limestone throws multifaceted challenges that need to be solved by applying scientific planning and emerging technologies. While assessing the mineral deposits, the degree of geological uncertainty observed during exploration should be minimized for proper planning and estimation of deposits. Both laterally and vertically, deposits should be explored in detail during geological investigations. As far as possible, the limestone deposit selected for exploitation should be geologically undisturbed and uniform in quality, particularly in hill areas. *The deposit type* or *form of deposit* should be a parametric norm for scientific planning in limestone deposit selection because this will facilitate exploitation.

The Environmental Management Plan (EMP) is based on an Environment Impact Assessment (EIA) of a mine and is a scientific and environmentally friendly way for mine reclamation and restoration. This scientific approach helps in learning the

A. K. Soni and P. Nema, *Limestone Mining in India*, Materials Horizons: From Nature to Nanomaterials, https://doi.org/10.1007/978-981-16-3560-1_6

pre- and post-mining scenarios of mine premises and surrounding area. Such an environmentally friendly approach applies to both day-to-day mining operation and advanced and predictive mine planning. In the case of an open cast mine of limestone, inadequate and non-scientific mine planning may result in long- and short-term losses. As a low-cost mineral, the margin of profit for limestone is small; hence, the risk cannot be taken for repeated planning and execution.

The use of *digital technologies* in the limestone mining industry can provide transparency and transformation in scientific planning. Adoption of such technological changes are therefore desirable and recommended for changing the working style through proper and continuous interaction between companies, employees, communities, and government. The importance of differential global positioning system (DGPS) surveys used to enhance exploration accuracy, reducing mining risk, and control illegal mining practices are now well established in India. Wherever its implementation is lacking, it should be adopted by both the implementing agencies (mine authorities) and the statutory agencies [Indian Bureau of Mines (IBM), Directorate General of Mines Safety (DGMS), Ministry of Mines, etc.)].

Carefully planned studies should be carried out to derive the full benefits of science, engineering, and management (see Box 6.1) for optimum exploitation with good conservation practices in limestone mining and industry. A rational approach to accelerate, modernize, and support the mining process from exploration to the decommissioning phase is needed to achieve the best possible deliverables, which in turn would prove beneficial to mining companies in terms of both long- and short-term business goals.

Box 6.1: Scientific Studies

- Detailed geological exploration, planning, and appraisal of exploration data and preparation of geological reports
- Feasibility studies for raw material availability, particularly *greenfield* as well as *brownfield* projects
- Opencast mine equipment selection and detailed engineering aspects
- Computer-aided mine planning, production scheduling, and deposit evaluation
- Productivity enhancement through process optimization for mines
- Performance audit of mining operations (technical) in limestone mines
- Environment Management Systems (EMS), including an environmental audit
- Performance evaluation of mining equipment
- Long-term mine sensitivity analysis based on radical constraints
- Analysis, evaluation, and assessment studies for drilling, blasting, explosives, slope, rock mechanics, geohydrological problems, and transportation optimization of ROM
- Troubleshooting in mining operations for sensitive areas, hill areas, etc
- Beneficiation of low or marginal grade limestone and blending studies

- Energy audit for mines and mining-related operations
- Metallurgical studies for mines and mining-specific processes for problems, such as lump formation, chocking of silos, corrosion, wear and tear of equipment such as, ball mills, vertical roller mills, and roller press used in the crushing and grinding of limestone.

6.2 Small-Scale Mining of Limestone

Small-scale mining, also referred to as *artisanal mining*, is equally dominant in limestone mining. In India, small-scale mines alone account for nearly 40% of the limestone production. Largely, these mines are unplanned and unscientifically developed. Therefore, while mining the limestone, due consideration must be given to the small-scale mining sector also. Two types of small mines exist in limestone mining areas.

1. Those developed naturally in course of decades of operation
2. Those preplanned and executed by some agency, association, or authority.

It is observed that category 1 small mines mentioned previously take a longer time to develop properly and have drawbacks in terms of the implementation/application of scientific approaches on account of financial restrains, whereas in the case of planned and executed small mines, it is convenient to carry out and shape the actual mining operation along modern lines in an eco-friendly manner.

Contrary to large-scale mines, small-scale mines are characterized by fewer requirements of reserves, implementation time, initial investment, and high employment potential. These mines require a moderate level of skills and infrastructure; hence, they take less time to develop.

Artisanal mining of limestone continues to employ conventional methods of exploitation. For such mines, low-cost engineering solutions are required. To a larger extent, this can be done by applying common scientific procedures. However, *best practice mining* and implementation of an *integrated approach* for planning and execution are the ultimate solutions. The following paragraphs present separate analyses of mining, the associated environmental problems encountered in small and large-scale mines, and practical mitigating measures.

In contrast to large-scale mines, small-scale mining of limestone needs consideration of local factors. In order to arrive at *best practices in mining* together with environmental practices, conventional methods and topographical intricacies of mining areas (i.e. hilly or plain areas) also should be taken into account. Some general points that address these issues along with possible solutions are listed here.

- Lack of environmental protection measures.
- Lack of synthesis of social and industrial needs for environmental protection.

- Mined out areas remain disturbed due to difficulties in reclamation and restoration through local means and measures.
- Deployment of equipment, machines, and accessories that are less damaging to the environment (i.e. environmentally friendly drill machine or crusher fitted with dust arrestor).
- Improper attention to biological reclamation of mined-out areas and dumpsites. (*Note*: Lower rate of survival of plants in mine areas creates hindrances for green revival.)
- Ad hoc mine planning in small-scale mines (i.e. lack of long-term and short-term plans).
- Disposal of mining waste is a big problem and challenge in small-scale hilly mines because the available area is limited.
- Unstable pit slopes in small mines may cause problems of land subsidence for which engineering measures are required.
- Vehicular movements on approach from roads give rise to dust generation and fugitive dust emissions. Spraying water on these roads along with their proper maintenance as environmentally friendly transportation practices can solve these problems.
- Other associated problems of mines in hilly areas (e.g. orientation and design of mine working faces, ugly scars or eyesores created due to mining, and the dumping of scree down the hills). These problems need to be addressed through scientific approaches and cost-effective measures.

Green belt development, *tree-clad approach roads*, and *tree barriers* placed along the haulage road and the lease periphery or on the outer limit of the quarry act as arresters for dust, fines, and flying fragments outside the mine. Such barriers also act as a curtain to cover ugly scars of the degraded land created due to mining and road development activity. A *community nursery* for planting and development within the limestone mining area is needed because it is useful from the production period to the decommissioning stage. This can be owned and maintained by a group of mines and should remain functional during the entire life of mine.

6.2.1 Wasteland or Pastureland

The damaging effect of mining on vegetation is well known, especially after mining is over or when the mining pit is abandoned. It is realized that in India *wasteland* is increasing and *pastureland* is shrinking (from 14,810 ha in 1966 to 10,258 in 2017). If mined-out land is grassed-over and converted into *pastureland* using suitable techniques, it will meet the growing demand of fodder for animals and increase milk output. This grassing-over of wasteland should be in addition to revegetation of the mined-out area as per the mine closure plan. These two measures would fulfil the requirement of restoring land to its near original shape and aesthetic look.

A mobile environmental monitoring laboratory for a group of small mines on a cost-sharing basis is suggested to monitor environmental parameters and manage them within limits during the operational stage of the mine(s). An internal wing of environmental management of mining operations for a group of mines is also desirable as this can do the job in a cost-effective manner [1].

On steep hills, inaccessible slopes, and mine dump sites, the spraying of seeds, soils, organic matter, binders or adhesives, and water in appropriate proportion to grow vegetation restores the excavated mining slope areas. This mixture is applied over the slopes at a pressure that allows proper seed embedment and germination.

6.2.2 Blasting Issues

Many times, the blasting operation in a single small mine becomes uneconomical because it requires full-fledged infrastructure including a dedicated blasting team, statutorily. To handle the blasting task economically, a group of small mines or a cluster of mines may join hands, especially for blast-related problems and issues of small-scale mines. Although small-scale manual mining does not involve the use of heavy machinery, there are cases where drilling and blasting can be eliminated. An alternative to the conventional drill and blast method is the use of *rippers*. Rippers can be used for limestone mining as limestone is a 'rippable' sedimentary rock (refer to Sect. 3.1.1) Using rippers in small mines or for a group consisting of more than one mine is economical and productive.

Instantaneous blasting or *blasting with delay detonators* in mines are executed in circuits. The biggest problem in small-scale mine blasting is that the delay detonators are either not easily available or (if available) their storage and cost become problematic for the small mine owner. Sometimes, their quality or specification (given by manufacturer) creates problems (meaning that they are not accurate in terms of delay). Any variation in milliseconds of delay may not result in the expected results. This is where a *programmable sequential blasting machine* (SBM) (8/16/24 or more delays) can be put to use for the exact number of delays required per circuit. To control the problem of flying rock and the damage to nearby infrastructures (i.e. cracks in buildings and houses), the use of an SBM (Fig. 6.1) as an environmentally friendly and economical piece of equipment could be promoted in small mines where it can be used collectively by a group of small mines. With the use of electric detonators or delay detonators in small mines, it is possible to contain ground vibrations and reduce detrimental effects of blasting.

6.2.3 Waste Disposal

Disposal of overburden or mining waste in mines is probably the most critical problem. This becomes more critical if restricted space is available for dumping.

Fig. 6.1 Programmable sequential blasting machine (SBM)

The method of waste disposal along a slope, which is commonly practised in hilly mines, is not the correct method and should be discouraged for waste disposal. It is desirable that local solutions (according to the mine topography) be adopted and implemented. Such solutions must take into account the cost involved. In addition, *waste disposal by backfilling* is the most appropriate method that can be adopted (Fig. 6.2). As the name of the method suggests, dumping of rejected materials or

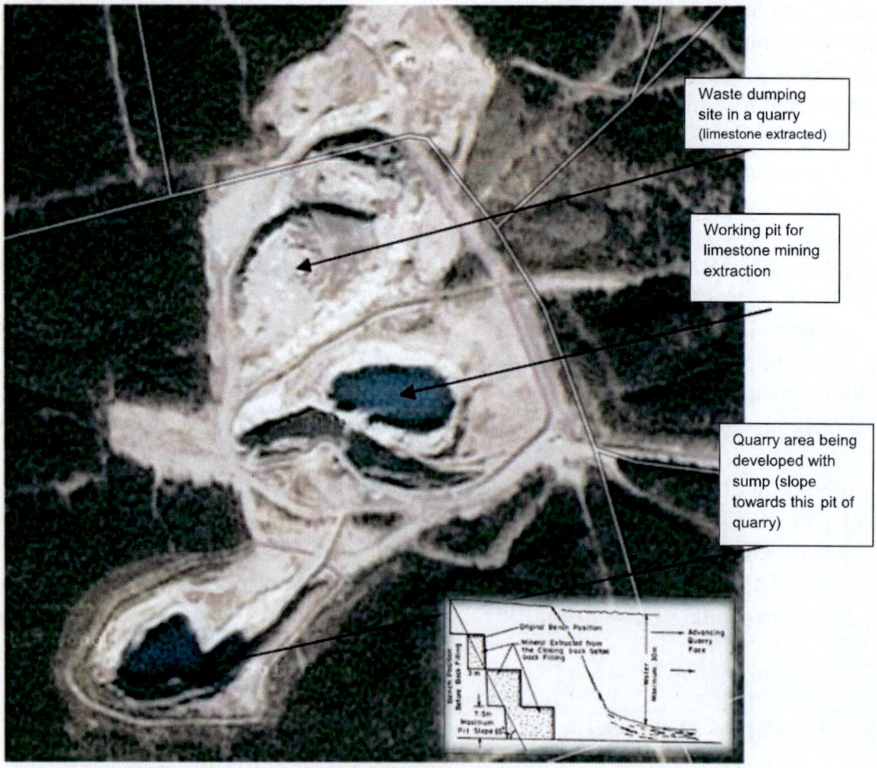

Fig. 6.2 Limestone quarry with waste disposal by backfilling

waste is to be done in the worked out area of the pit; thus, the waste is contained in the mining premises only. The method of backfilling is speedy and cost-effective because the transportation distance is less.

For waste dumping, transportation can be accomplished using a conventional truck or dumper system because the other systems of transportation (like an aerial ropeway or belt conveyor systems) are costlier and require essential prerequisites, such as the proper location of the waste dumping site, etc. Meticulous planning and layout development—right at the initial planning stage—are needed for the transportation of waste.

Another method suitable for waste disposal is the method of *benching* (i.e. *waste fill construction in lifts*), which can be chosen for waste dumping if the topographical features allow. In this method, benches or steps of about 1.5–2 m in height are formed on the hillside. An approach road is required to be developed up to the dumping face. This situation arises wherever limited space is available for the disposal of waste (e.g. hills). As a general practice, the waste or overburden in mines is handled using this method, creating heaps in bench forms. For hilly areas, this dumping method is quite suitable. Walls made of stone obtained from the valley and along the slope should be erected in hilly terrain. By doing this, scree flow on slopes is restricted, and a typical mine waste disposal system in the limited area is maintained along the contour of the hill. Both methods mentioned here should be planned systematically so that environmental protection target is achieved to the maximum possible extent.

The problem of water pollution (i.e. discharge of effluents and scree flowing in the water channel) is caused by the limestone mining in both plain and hill areas. Diverted surface water enters into the mine from adjoining areas together with the drainage water from the mine. This causes the water pollution in water bodies or water channels around the mining areas and needs to be regulated properly. Solutions include the following measures.

- Diverted surface water entering into the mine from the adjoining areas needs to be restricted as per the ground elevations (RL).
- Sumps at the lowest RL of the mine and drains along individual benches discharging into the sump could be provided for maintaining drainage within the mine. To prevent soil, waste, and debris from getting into this drainage area, gravel or sand packed pits (Fig. 6.3) can be constructed at number of places along the periphery to arrest the choking of surface drains.
- Construction of a series of toe walls (in hilly areas) using local materials (i.e. rocks and rubble from the mine).
- Planting shrubs, grasses, and saplings in the area to help prevent soil erosion.

6.2.4 Water Flow and Filtering

To prevent the flow of scree in the water channels, *check dams* also can be built at regular intervals. Provision of such structures reduces the velocity of water and prevents the scree flow.

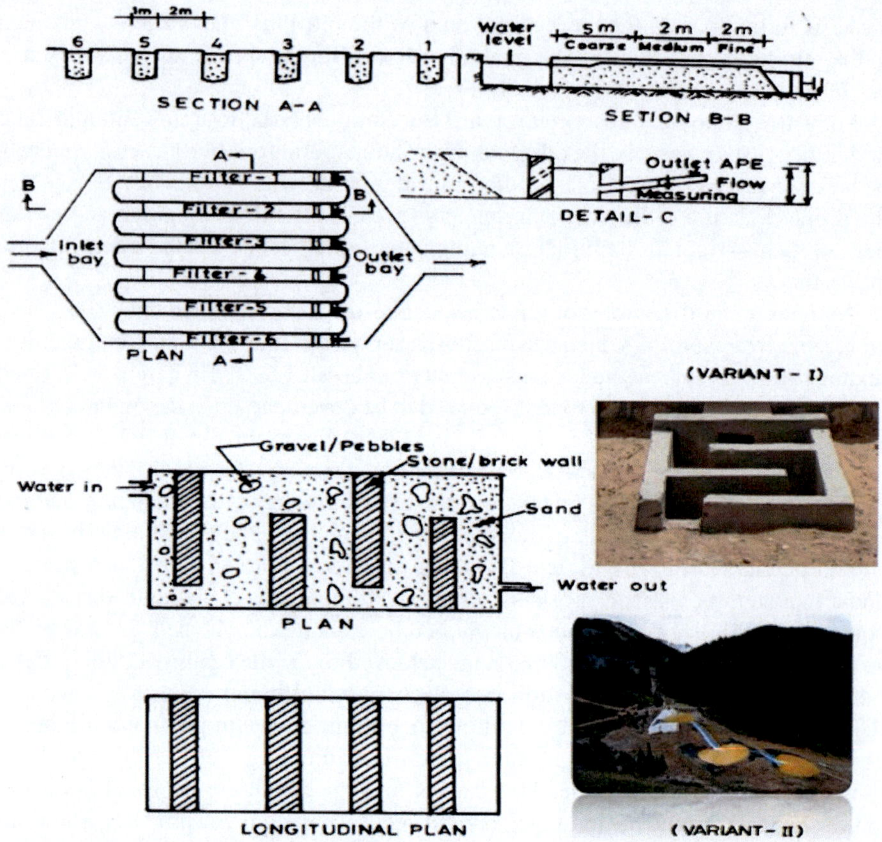

Fig. 6.3 Gravel and sand packed pits. *Source* [8]

On steeper slopes (with angles more than 25°), *check walls* and *check dams* arranged with filtering material (Fig. 6.3) should be constructed. To construct such structures, 25 mm holes of 01–1.5 m depth are drilled at 60–100 cm intervals across the watercourse in multiple rows. In between the rows, a 50–60 cm distance may be kept. In these holes, steel rods of 2–4.5 m length are grouted. These are then welded in a criss-cross or diamond-shaped network 10 mm in size *tor* steel rods set at 60 cm apart. The space between the tor steel row is filled with rubble and boulders. On the top, all of the rows of tor steel rods are welded together to make a single unit. A number of steel pipes (150 mm diameter in size) are also fixed in place, as shown in Fig. 6.4. These dams or walls are filled with sand, gravel, and boulders to act as a filter for incoming water.

Check dams also can be built easily from the scrap steel. These steel structures are cheaper than the wall type of structure and are strong enough to withstand the impact of rolling boulders and to arrest the build-up of silt in water.

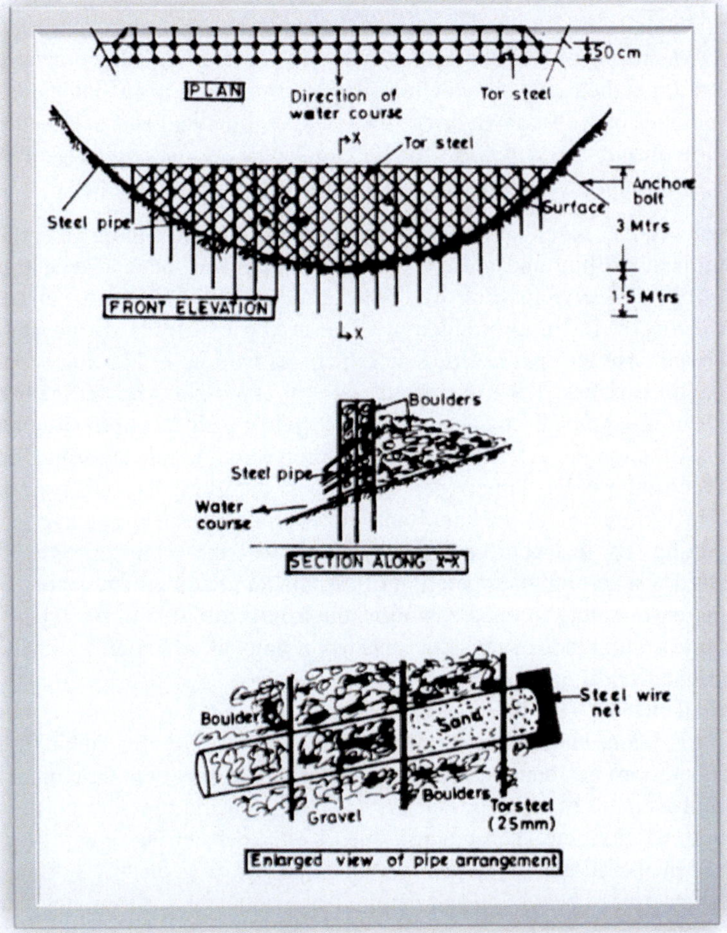

Fig. 6.4 Check walls and check dams with filtering arrangements. *Source* [8]

6.3 Improving Mine Productivity

Limestone mining as an industrial activity is very hardy, dirty, and repetitive. Productivity of any limestone mine in a commercial sense is a combination of two points.

1. Optimizing total cost of production with safe mine operation
2. Controlling the size of produced material as per requirement (e.g. finer size for cement manufacturing and coarser size for aggregates in construction, road, and infrastructure project applications.

If the productivity of a mine is improved, the profitability goes up. Improvement in the productivity of limestone mines is possible by *mechanized extraction* and *mechanical handling* of ROM, which is bulky. To make the existing practices easier, the best option is the use of suitable machines for mining. Labour-intensive excavation has number of well-known drawbacks that are outdated and not needed. Also, the adoption of *zero-waste mining practices* and *value addition approaches* has been advocated to ensure productive and sustainable mine operation [2].

- **Modern Tools**: To enhance the scope for sustainable mining operation, unit operation of drilling and blasting can be modernized. Several new techniques of *controlled blasting* for improved *powder factor* (fragmentation) can be helpful in advancing overall mine productivity. This can be done by acquiring site-specific knowledge through a parametric scientific investigation for blasting to solve critical technical issues. The use of better quality explosives and detonators, which have desired technical characteristics, also will result in improved energy and velocity of detonation (VOD) that are required for an efficient production blast.
- **GPS Technology**: For larger open cast mines, the truck dispatch system should use *GPS technology* as an important and useful tool for productive and effective mining. By this approach, the cycle time of the mining activities can be brought down and remote operation of machine and equipment is possible. This also increases safety in open cast mine environments. It is in the interest of the limestone mining industry to adopt this technology as an Indian vision 2030 goal to improve overall mine productivity.
- **Grade Control**: The broad area of *grade control* for raw material quality, plant utilization, and surveying of the mine and mining area (including equipment operation) has innumerable scope for cost optimization, time management, and improvement of mine productivity. The increasing pressure on production, cost optimization, etc., can be cushioned by further cutting down the expenses incurred in operational processes of open cast mines [e.g. electricity consumption through energy conservation and maintenance of heavy earth-moving machinery (HEMM)] on a daily basis.

Efficient transportation systems can make an existing mine more profitable as the overall cost of production can be brought down substantially in the long run. Most mines use road transportation with trucks for hauling and dumping for the practical reason of flexibility. Substituting this form of ROM and OB transport with a continuous mode of transport [e.g. belt transportation and slope hoisting arrangements (Fig. 6.5)] will contribute to better productivity and long-term cost savings. Therefore, improvements and cost-effectiveness are achieved by adopting environmentally friendly and comparatively economical modes for transportation.

It is beyond doubt that better mine planning enhances productivity. Under the existing practices, the permissible height of benches in small open mines is only 1–1.5 m, whereas 10–20 m is needed in the case of large mines. The bench widths are planned according to the bench heights. However, to maintain the targeted production, sufficient bench heights up to 4–6 m are needed. Such dimensions of benches can be easily planned, but the statutory requirements of mining laws do not permit more

Fig. 6.5 Slope hoisting arrangements. *Source* Siemag Tecberg mining technology

height. This gap in mining law needs to be addressed and rectified on a case-to-case basis for improved productivity from mines. If this point is taken care at the development stage of the mine or made flexible with regard to small mine site conditions, improvement in productivity of the company coupled with better mining conditions can be achieved. To reshape the future of limestone mining in India, the use of modern machines—like the unmanned aerial vehicle (UAV) known commonly as a drone—as tool for the traditional survey output is suggested. The UAV use can minimize the labour required and time consumed for prospecting, exploration, construction, development, production, and reclamation in the mining life cycle (Fig. 6.6). This device also has proved to be a boon for use in inaccessible areas.

Limestone availability for indigenous requirements should be ensured from internal resources, so there would be little or no export needed of this low-value

Fig. 6.6 Unmanned aerial vehicle. (UAV) (drones) for mining applications

mineral. Hence, every effort should be made to bridge the gap between supply and demand. A *detailed exploration* of the limestone reserve is required for an accurate assessment of national need. If this reserve position is known, it can be very useful to bridge the supply–demand gap. The United Nations Framework Classification (UNFC), being adopted by concerned institutions for project funding, has relevance to the Indian limestone mining industry. It needs to be implemented and periodically reviewed.

It is extremely necessary that all mechanical equipment used for handling limestone and waste rocks be updated and run efficiently. All plant and mining machines, HEMMs, crushers, etc., have tremendous potential for cost savings in addition to improving the overall economic benefits of limestone production. A variety of practical experiences have been gathered during the actual use of machines in mines and mining operations. Based on these experiences and available literature and research, some aspects related to both mechanical and electrical use that resulted in enhanced productivity and cost savings are highlighted in Box 6.2.

Experience has shown that the maintenance of HEMMs; overseeing the optimum and correct use of prime movers (engines); checking the condition of engine oil and the oil change cycle period; and monitoring equipment such as pumps can eventually bring down the cost per tonne of limestone. *Energy savings* and *quality maintenance* are areas where saving of money is possible in limestone mines.

Box 6.2: Mine Productivity and Plant Equipment and Machinery

Limestone Mine Crushers

(a) Hardened steel picks attached at the front end of the bucket of hydraulic excavators used in mines for limestone handling (also called *tooth points*) are dismantled many times and can accidentally fall into the crusher. This will cause severe damage to the crusher, resulting in a major breakdown. To overcome this difficulty, tooth points are welded in place on the bucket. This allows a maximum use of the tooth points, achieving a life of 4500 h (in place of the previous 1500 h). This welding alone can save up to Rs. 0.15 million per year.

(b) Rebuilding the blow bars in an impact crusher and conducting a vibration analysis should be done every quarter in order to diagnose the defects at the necessary time. Cost savings and also achieved by using state-of-the-art equipment, machinery, and technology.

Routine Condition Monitoring of Engines and Transmission Systems of Mining Machines

The condition of engine oils of deployed machineries should be monitored before a scheduled oil change. This is done using lubricant oil test kit. Wherever possible, an oil change period can be adjusted by 50 h, resulting in an overall savings of Rs. 0.21 million in a three-year period. In addition, a saving of Rs. 0.141 million in oil and 0.067 million in filters is made.

Savings in Transportation Costs

(a) The maintenance of dumpers and other trucks deployed for the transportation of limestone using a *centralized greasing system* saves considerable cost. Savings of Rs. 1.6 million per annum coupled with reduction of manpower, breakdowns, etc., can be made.
(b) If mine roads—approach as well haulage—are maintained correctly, the annual and monthly consumption of fuel (diesel), tyres, and replacement of brake drums are reduced. This will result in saving of lakhs of rupees per year. For proper maintenance of haul roads and mine roads, deployment of a *road grader* is suggested. Grading the roads will result in less breakdown of HEMM'S and other vehicles and will reduce the overall cost of maintenance. The average life of dumper tyres is 5000 h. However, with improved road conditions, 7000–8000 h tyre life can be achieved.
(c) The cost of using a tyre-handler is more than Rs. 30 lakhs. However, it can be installed at site at a cost of 8–10 lakhs, making dumper tyres handling operations (i.e. their transportation, fitting, etc.) more convenient. This also uses less labour and makes completion of such maintenance work faster.

Having a locally developed *audio-visual horn system* at a plant or mine workshop for heavy earth-moving machinery (HEMM) has saved considerable amount on the maintenance head. Also, a locally developed *water injection system* for 100 mm down the hole (DTH) drills of mines that are deployed for breaking rocks resulted in further reduction of costs for production.

Note Cost figures have been taken from Sarkar [3] and are indicative only. Figures may vary as per the prevalent market rate.

A rough estimate of monetary gains is highlighted in Table 6.1 for the plant, equipment, and machinery used in mines. In a mechanized limestone mine, these machines are necessary for the bulk solid handling. Hence, periodical upgrades and timely maintenance are very important for improving the mine productivity and the cost savings in mining.

6.4 Long-Term and Short-Term Trends

Mining is a process of mineral extraction from the earth (ore body), and we need a fine balance between the exploitation of natural mineral resources and socioeconomic development of society. In this regard, both long-term and short-term market

Table 6.1 Monetary gains at a glance*

S. No	Item description	Gain/year (Rs. in millions)
1	Lubricant oil for machineries	0.07
2	Overhauling of engine and transmission	0.80
3	Rebuilding of tooth points of excavators	0.15
4	Audio-visual reverse horn system	0.095
5	Brake drums	0.35
6	Rebuilding of blow bars	0.20
	Total monetary gain per annum	1.665

*Figures indicated are approximate and indicative and are subjected to variation as per the prevalent market trend

trends and the latest global practices are to be followed or adopted. When the limestone mining industry is being operated by private sector participation (nearly all cement manufacturers are private companies), *eco-friendly* and *green mining initiatives* should be incorporated in the business philosophy of all such mining companies irrespective of their size and production capacity.

Planning and development are the two important stages of mining that are interwoven. In the case of any mine—small or big—two types of developmental planning are essential: long-term and short-term plannings. Long-term planning is required for the reclamation of a mined-out area and should be carried out in a phased manner right from the inception. This will enable the design of the mine in a systematic manner. The planning for transportation, waste management, etc., should be done in the longer perspective with a short-term focus. It is essential that short-term plans be commensurate with the long-term objectives. Both long-term and short-term gains (e.g. *biodiversity initiatives*) for the better environment of mining areas are the anticipated results in future.

Long-term technological solutions for the design of an ultimate pit slope, with overall stability of the pit slope using modern instrumentation, are a current trend. Slope stability and management of pit slopes through *back analysis*, including failure prediction for mine slopes, is yet another robust and reliable long-term solution for carrying out sustainable mining [4].

With the recent development in the mining of low-grade ore and new techniques of exploration, it is possible to bring some more limestone areas on the mineral map of India in the near future, thereby attracting investors. Many areas in different Indian states, such as the newly created Telangana state, where limestone deposits have been less explored, could be the beneficiaries of such attempts. Efforts in this direction will pave the way for the development of more limestone mineral bearing areas. Use of NextGen technologies would be helpful in maintaining green limestone mines in different areas.

6.5 Local Solutions and Benefits

Mining activity accentuates the twin problems of pollution (air and water) and land degradation. The conventional mode of mining and new ways of limestone handling (or a combination of both together) can provide beneficial solutions for limestone excavation. Many times, global problems are solved with local solutions. With this perspective in mind, some pertinent issues are discussed here (e.g. crushing, transportation, cost of operation, and enhancement of mineral quality.

In-pit Crushing

Recent experiences have shown that, at certain pit depth, in-pit crushing and belt conveying show significant cost savings. Figure 6.7 shows an in-pit crusher with a belt conveying system. Cost comparison (for capital cost and operation cost) between truck transportation and in-pit crushing with a belt conveyor in relation to pit depth for the mines in a plain area is shown in Fig. 6.8. The relative cost figures are based on typical installations where the breakeven point would vary from mine to mine. However, as a general rule, the deeper the pit and the larger the daily tonnage mined (production), the more beneficial the belt conveying system will be [5]. It is seen that local in-pit crushing and conveying systems developed at a mine workshop work quite effectively.

Transportation of Mineral and Overburden

Transportation of mineral and overburden (OB) is a major area where local cost-effective solutions can be implemented. The use of gravity for mineral and OB transport in hill areas offers good opportunities [e.g. shaft–adit combination (SAC) method, gravity operated skips (GOSs) on guided rails, ore inclines, etc.] [1].

Fig. 6.7 In-pit crusher and conveying system

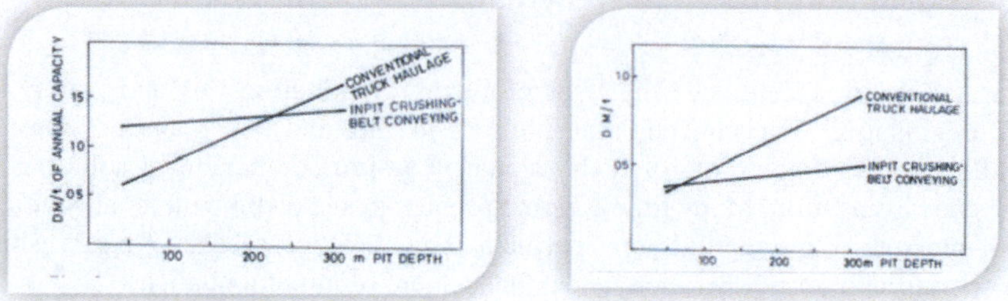

Capital Costs *Operating Costs*

Fig. 6.8 Cost comparison between truck transportation and in-pit crushing with a belt conveyor for reducing capital and operation costs in relation to pit depth

Further, transportation by surface roads generate a huge amount of dust by dump truck movement and is a major source of air pollution. From an economical angle, the running cost of dumpers—particularly in hills—is higher than in the plains. A comparison of transportation cost to the cost of other unit operations (such as blasting and drilling) to produce one tonne of limestone indicates that for normal plain areas and in hill areas there exist a wide gap [1]. This great difference can be closed by using local measures that are cost-effective in implementation.

Environmental Management

- In limestone mines, dust is a major problem, and its suppression consumes a lot of money and energy, reduces work efficiency, and creates health hazards. Mine dust not only affects mine employees but also affects the neighbouring community. Hence, to reduce the negative effects of dust emissions, thereby creating a safe and healthy environment, a dust suppression system has been designed in-house and installed at the crusher plant at Gujarat Cement Works (GCW) in Kovaya (see Fig. 3.2 in Sect. 3.1). This demonstrates that local solutions are readily available for implementation.
- Producing less mining waste will make its management easier. Hence, a philosophy for zero waste or generation of less waste works well for limestone mining and its judicious use as a raw material.
- Progressive restoration of the quality of land using *biological reclamation* natural methods is a local, permanent, and cost-effective method. Biological reclamation allows for green mining practices during the operational phase of a limestone mine.
- The use of local available materials in a mining area for the construction of check dams, groundwater recharge structures, and siltation ponds or basins is an economical way of water management.
- Limestone is a carbonate sedimentary rock consisting of calcium carbonate as the main constituent and magnesium carbonate as a secondary component. The impacts of limestone mining on the environment are many [6]. The impact on

water sources being one of the major concerns. Open cast mining of limestone has an adverse impact on the quality of water in the nearby areas and can result in higher levels of pH, total dissolved solids (TDS), total hardness, alkalinity, as well as calcium and sulphate concentration. It may also result in acid mine drainage, if a sulphur-bearing stratum is in the vicinity of limestone deposit, which can release toxins into the nearby water bodies. Consequently, surface and groundwater quality of the existing water sources can become impaired and rendered unfit for human consumption. This creates a serious problem for the local population [7].

To solve the problem, a twofold management strategy needs to be adopted such as providing appropriate treatment to wastewater generated from mining before discharge into surface and groundwater by making this water reusable for miscellaneous purposes as well as practising water conservation measures that, in turn, would reduce demand pressure on available freshwater sources. As mine pits are dewatered, huge volumes of water are generated, which can be utilized after desilting and removal of suspended solids for dust suppression and plantation in the mine area. A properly planned and designed network of drains with settling basins will prevent the flow of dust and fragmented material to nearby drainage. The rainwater accumulated in mined-out pits will recharge groundwater and maintain the groundwater table of the region.

For making the water reusable for other purposes, conventional water treatment techniques can be applied depending upon the intended use. The conventional water treatment methods comprise presettling, pH correction, coagulation-flocculation followed by sedimentation, filtration, and disinfection. Not all the above treatment steps are needed for the water of every mine, but depending on the individual case, they can be applied selectively. The basic purpose of water treatment is to make the water safe by ensuring that it has a balanced pH and is free of inorganic and organic impurities including pathogens. A few more treatment steps may also have to be added to conventional treatment to make water fit for human consumption. Currently, many advanced water treatment technologies are available (e.g. activated carbon filtration, ultra-violate radiation, reverse osmosis, etc.). Some are more effective than others and can be applied on a case-by-case basis.

Mineral Improvement and Involved Processing

Limestone as a mineral does not requires much improvement or processing like metallic ores. However, challenges in limestone quality improvement using advanced ore processing techniques are surely to come in future. Since limestone is available in abundance, its quality has not drawn much attention and is largely ignored. It may be noted further that the generation of high-purity or high-grade calcium from limestone for various uses in outer space applications may be required in future by the industry, and this can be derived only through limestone processing.

6.6 Technovations

Limestone mining in India pertains to its excavation and properties for various uses. Limestone, being the most coveted primary raw material ingredient for the cement industry, occupies an important rank as the most abundant low-value mineral category. The current trend of limestone mining in the Indian mineral industry harnesses emerging technologies to redefine and reinvent solutions that suit production optimization and reduce costs towards bulk limestone production.

Excavation, digging, loading, transportation, crushing, and grinding at mine pit head are all essential processes of limestone production that do not involve rocket science or precision. Mining certainly requires scientific inputs for productive output, which is required by the industry. As a result, mining draws the attention and interest of the science and engineering community. It is mainly the costs involved in the complete process—from concept to execution—that need to be curtailed so the production can be optimized, possibly through mechanization, with the aid of the industrial Internet of Things (IIOTs), artificial intelligence (AI), virtual imaging, and many other cutting-edge technologies. Sometimes, the conventional and traditional methods also become cost-effective and have an edge over modern technologies. Depending on the realities encountered in the field, one has to choose the correct course of action for a specific mine for the best results.

Transforming limestone mining into a business with safe and productive practices is indeed a challenging task with various risk factors involved. However, the evolution of digital technologies, automation, and advanced equipment has allowed the limestone mining sector to grow not only in India but also in other parts of world. Evidence suggests that a *productive, cost-effective*, and *eco-friendly mining system* can be developed through the approaches provided in this book. Accordingly, efforts have been made here to cover all of the necessary details of mining in general (and limestone mining in particular) to present the existing scenario. It is hoped that the given descriptions and case studies will be helpful both for the planner and executors of the mining industry.

India is the second largest cement-producing country in the world with a distinction of operating more than 500 plants of varying capacity and technologies. Therefore, limestone is its main extractable raw material, and its mining has the following technology options for improvement (Table 6.2) with a focus on the cement industry.

An elaboration of the listed technologies in Table 6.2 requires a lot of discussion. Therefore, for the sake of brevity, it is left to the reader's wisdom and discretion to select, research, and apply, one or more of these options, depending on a specific situation or case. Obviously, technology continues to upgrade with time, and so is the process of limestone mining.

Responsible mining is an intricately framed new term used to explain and create safe mining zones from commencement to decommissioning of mining operations. Starting from conceptualization, in-principle statutory clearances (e.g. granting a mining lease, environmental clearance for operating mine, providing safety clearances) can be simplified to obtain a desired ROM production. This helps curtail

Table 6.2 Technology options for limestone mining

S. No	Area	Technology options	
		Existing	Modern as per current global trends
1	Mining or excavation of limestone	Conventional or manual	Fully mechanized/semi mechanized
2	ROM and material handling	Conventional	computer-aided material handling equipments (e.g. satellite-controlled truck dispatch systems, reclaimers, and GPS-based remotely operated dump trucks
3	Crushing and grinding	• Jaw crusher, impact crusher, or ball mill with or without conventional classifiers • Two-stage crushing and grinding	• In-pit crushing and conveying • Vertical roller mill (VRM), roller press and dynamic classifiers • Single-stage crushing and grinding
4	Transportation and conveying material (away from mine to utilization point)	Dumpers, tippers, and other trucks	Continuous mode of transport using belt conveyors, pipe conveyors, or high-angled conveyors

the delay and costs of beginning limestone mining operations. As an alternative to open-pit mining (or open cast stripping), which is in vogue in India for limestone extraction, there lies the possibility of underground limestone mining too. The possibility of underground limestone mining in future is immense, but it has yet to be developed fully in India. It is feasible that other factors may turn in favour of underground limestone mining, as it is used in other parts of world.

In summary and with respect to modern technological options, *energy*, *environment*, and *employment* are three key areas surrounding the growth of the limestone mining sector. These areas should be used to improve the Indian mining scenario and promote positive progress. Also, the methods of extraction and their impact on mining operations should incur benefits for economical and societal gains as well as promote healthy environmental conditions for the local communities of limestone mining areas. Corporate social responsibility (CSR) activities and infrastructure building are tools for the implementation and development of desired benefits in limestone mining, including helping with the responsibility of sustainable development in limestone mining areas.

References

1. Soni AK (1997) Integrated strategy for development and exploitation of natural mineral resources of the ecologically fragile area, Ph.D. Thesis (Unpublished), Indian School of Mines (IIT-ISM), Dhanbad, September, p 238
2. UltraTech Cement (2017) Bullish on growth building a sustainable future. Sustainability report, 2016–17, UltraTech Cement. www.ultratechcement.coms, p 113
3. Sarkar SK (1995) Routine condition monitoring in mining. Oxford and IBH Publishing Company Private Limited, New Delhi, p 147
4. Jhanwar JC, Buragohain J, Sangode AG, Patel P (2019) Geotechnical study to evaluate pit slope design of a limestone mine for safe and optimum mineral extraction. In: Singh PK et al (eds) Proceedings international conference and exhibition on energy and environment: challenges and opportunities (ENCO-2019), Feb, New Delhi, pp 934–941
5. Kutschera SA (1984) Planning aspects for the application of continuous transport systems in hard rock open pit mines. Int J Bulk Solid Handling 4(3):609–613
6. Ganapathi H, Phukan M (2020) Environmental Hazards of limestone mining and adaptive practices for environment management plan. In: Singh R, Shukla P, Singh P (eds) Environmental processes and management; water science and technology library, vol 91. Springer, Berlin, pp 121–134. https://doi.org/10.1007/978-3-030-38152-3_8
7. Lamare RE, Singh OP (2014) Limestone mining and its environmental implications in Meghalaya, India. Int J Environ Sci 3(5):13–20
8. Soni A.K. (2017) Mining in the Himalayas—An Integrated Strategy, CRC Press / Taylor & Francis, p. 225

Chapter 7
Epilogue

The Indian mining industry offers an enormous potential of core raw materials for cement production, which include limestone. Cement manufacturing also requires other minerals in addition to limestone, namely gypsum, coal, and kaolin (china clay) in varying amounts, and these account for a total share of about 5–10%. Gypsum is the most essential ingredient of finished cement that is also obtained from mining [1]. Though useful, the handling of gypsum is not so easy as it is a hygroscopic material and very sticky in nature [2]. All of these mineral resources are in abundance and produced indigenously at a very economical cost from mines. Limestone, gypsum, and coal as raw materials are anchored on the demand, supply, and pricing. Depending on the grade of cement, the gypsum percentage varies. According to an estimate, about 1.4–1.5 tonne of limestone, 180–250 kg of coal, and 2–3% of gypsum are required to produce one tonne of cement.

Analysis of various aspects of limestone mining in India make it evident that the demand from the cement sector, for which limestone is the major basic raw material, will continue to grow at a rapid pace. This will ultimately result in the commensurate production of limestone and investments in the limestone mining sector. Needless to say, such opportunities should be availed by the industry and the accrued financial gains be shared with society.

The policies regulating mineral commodities and their concerned industries are considered key and important because they influence production as well as demand and the same holds true for cement industry also. The industrial demand benefits society at large because of its potential to generate several opportunities (e.g., employment, infrastructure development, and healthcare). It is also apparent that all of these measures accelerate and contribute to increase in the gross domestic product (GDP), which is an indicator of nation's progress. The Indian cement industry (Fig. 7.1) is the second largest industry in the world and is poised to grow to 550 MTPA (million tonne per annum) capacity by 2025 (up from 502 MTPA in 2018). limestone mining and the world cement demand per unit of GDP for developing countries (Fig. 7.2) for the period between 2015 and 2025 (estimated) show that China and India account for around half of the global cement production.

A. K. Soni and P. Nema, *Limestone Mining in India*, Materials Horizons: From Nature to Nanomaterials, https://doi.org/10.1007/978-981-16-3560-1_7

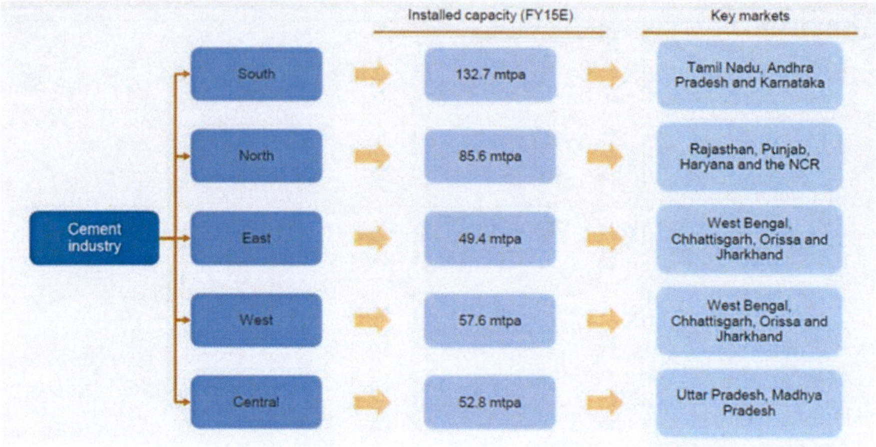

Fig. 7.1 Indian cement industry. *Source* [5] www.ibef.org (February, 2017)

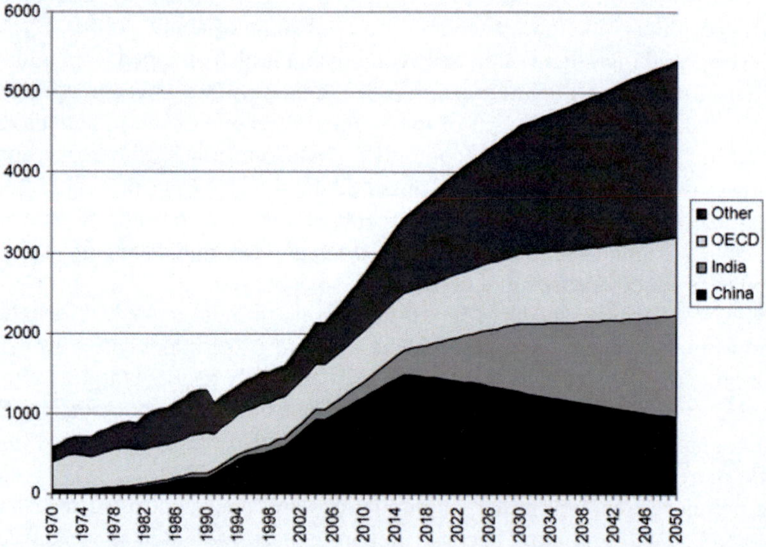

Fig. 7.2 Global cement demand by region and country (1970–2050). *Source* Taylor et al. [3]

With the Indian government targeting for 7–8% GDP growth and the revival of infrastructure and real estate sectors, it is expected that government spending for industrial and related infrastructure development in rural areas will be more and consequent mineral demand will continue to exist for a long time. Taking this into account, it is noted that north-east India has enough unexploited reserves with assured long-term availability to meet present as well as future domestic demands of the region adequately. Domestic demand for limestone in the north-east region is expected to remain strong, and this region will have attractive long-term industrial growth. In the next 10 years, India as a whole can become the main exporter of clinker and grey cement to the Middle East, Africa, and other developing countries of the world. Cement plants near the ports, for instance the plants located in Gujarat and Andhra Pradesh, will have an added advantage for exports (https://www.ibef.org/industry/cement-india.aspx). Limestone consumption centres in the interior of the country, along coast, or in virgin (new) areas will have an advantage but will face stiff competition in future.

A review of the limestone-consuming private sector [4] has indicated that the technological options, industry up-grades, and continual improvements are incorporated as a routine exercise in mines. Compared to mines of other minerals, most of the limestone mines are progressively and aggressively implementing clean, green, and planned practices. Further, positive growth in terms of socioeconomic development (i.e., quality of life improvement) will be the hallmark for limestone mine areas, which is a good sign. The north-eastern states of India (as mentioned previously) are likely to be the new and important growing markets for cement companies creating increasing demand thereby boosting the, limestone production in the future.

In Indian mines, limestone mining operations, such as excavation, ROM handling, crushing, grinding and transportation or conveying materials, have ample scope for improvement through innovative and emerging technologies skill improvement, and better use of human resources. Over time, this sector will achieve optimal economical solutions in terms of cost of production per ton, increased capacity etc., In the mining industry, a realization has developed that the energy consumption in crushing, grinding, ROM handling, transportation, and conveying of limestone is very high and needs to be brought down. Therefore, electrical energy consumption in mines is yet another area where tremendous potential for improvement lies. With the implementation of energy-saving measures, more economocal prodution is also possible which in turn will lead to overall socio-economic development of the country in an environmentally sustainable manner.

Thus, limestone, a non-metallic mineral of major mineral category,is one of the major contributor to the Indian exchequer in terms of revenue (value), mineral royalty, District Mineral Foundation (DMF) trust funds, and National Mineral Exploration Trust (NMET) fund.

From an environmental perspective, adoption of cleaner technologies and commitment towards a cleaner environment (e.g.reducing CO_2 emission from mining operations) should be the target and responsibility of each limestone mine owner—now and in the future in order to ensure environment-friendly and sustainable mining.

In this book, a detailed analysis of various aspects of the Indian limestone mining industry has been presented. Attempts also have been made to highlight areas of possible improvements and up-grades in mining processes with recourse to newer technologies and emerging scientific tools. It is emphasized that mining minerals in particular and any industrial production in general have to be environmentally friendly. It is then—and only then—that the goal of sustainable development can be achieved and society at large can enjoy the benefits—be it from industrial expansion or the overall socioeconomic growth of the country.

References

1. Khadilkar S (2019) The role of gypsum in cement. Indian cement review (ICR), Nov, pp 48–52
2. Damle V (2019) Gypsum, the speed breaker. Indian cement review (ICR), Nov, p 42
3. Taylor M, Tam C, Gielen D (2006) Energy efficiency and CO_2 emissions from the global cement industry. IEA-WBCSD workshop, Energy Technology Division, International Energy Agency (IEA), Paris, Sept, p 12
4. IMYB (2019) Indian mineral year book (IMYB) 2019, Part-III: limestone and other calcareous material, ministry of mines, Indian Bureau of Mines (IBM), Nagpur, Government of India, Advance release, 58th edn, July-2020, pp 18–1 to 18–21.
5. IBEF (2017) India brand equity foundation (IBEF). www.ibef.org/industry/cement-india.aspx. Accessed in Feb 2020

Annexure A

Limestone Belts of India

(Referenced in Sect. 1.4.1)

A. K. Soni and P. Nema, *Limestone Mining in India*, Materials Horizons: From Nature to Nanomaterials, https://doi.org/10.1007/978-981-16-3560-1

S. no.	Limestone origin	Name of state	Locality	Cement plants and cement manufacturing	Remarks
1	Archaean Group (oldest)	Rajasthan	Kheri, Mandaria, and Drauli of Udaipur district	13 operational cement plants	Dolomitic limestone
		Karnataka	Nagargali-dandeli of Belgaum district		Dharwar period
			Ballur, Arakere, Shankaragudda, Bikonabhalli, Joladahalu, Rangapur, Malliger, Bhadigund, and Gangur Urbani of Shimoga district		
			Vobalpur, Dodguni, Chiknayakanhalli, Huliyur, and Nelanhalli of Tumkur district		
			Kanchipur, Kudure-Kanive State RF, and Javangondanahalli of Chitradurga district		
		Odisha	Hathibari, Gatitanagar, Tikamotoli, Pahartoli, Lanjiberna, Ludhukutoli, Birmitrapur, Khatkurbahal, and Purkapola of Sundergarh district		
		Tamil Nadu	Ramayyanpatti and Talaiyutu of Tirunelveli district		Crystalline limestone of cement grade
			Pandalkudi, Palavanatam, and Alangulam of Ramnathapuram district		-do-
			Gajilampatti, Kullampatti, Dholipatti, Muthampatti, Devaramalai, and Ayyampalayam of Madurai-Trichy district		-do-

(continued)

(continued)

S. no.	Limestone origin	Name of state	Locality	Cement plants and cement manufacturing	Remarks
			Madukkurai and Ettimadai of Coimbatore district	ACC Madukkarai	-do-
			Sankaridurg, Malappalayam, and Pudupalayam of Salem district		-do-
		Bihar	Bundu-Basaria, Kurkutta-Religara, and Hosirbachra-Bunduray of Hazaribagh district		Crystalline limestone
		Himachal Pradesh	Naura, Bhangari, and Jarag of Shimla district		Jutogh series
2	Cuddapah system	Andhra Pradesh	Cuddapah basin	Bagalkot (Karnataka) Banas (Rajasthan)	New large-capacity cement plants added that uses Cuddapah limestone. Came recently in Rajasthan, Maharashtra and Karnataka states
			Asifabad and Tandur		Pakhal series
			Adilabad		Penganga Basin in the Pranhita Valley
		Bihar	Singhbhum (Chaibasa and Jagannathpur)		Kolhan series
		Chhattisgarh	Bhopalpatnam of Bastar district		
		Karnataka	Hoshatti, Mannapur, Aurad, Thimmapur, Yadwad, and Manami in Belgaum district		

(continued)

(continued)

S. no.	Limestone origin	Name of state	Locality	Cement plants and cement manufacturing	Remarks
			Kaladgi, Yendigeri, Hira Sellikere, Chandapur, Khajjidoni, Varchagal, and Katgeri of Bijapur district		
		Gujarat	Khunia, Pasuval, Diwania, and Karamudi district Banaskantha		Crystalline limestone of cement grade
3	Vindhyan system	Chattisgarh	Kosamdih Mohtara, Akaltara, and Gatora of Bilaspur district		
			Birgahami, Darrabhatta (Pali), Sarkhon, Karmandi, and Khaira of Raipur and Balodabazar district		Chhattisgarh—Raipur Basin limestone
			Arjuni, Bhanpuri, Ghoda (Gundrdehi), Khuteri, Limora, Sikosa, Sukhri, and Kirgi of Durg district		Charmuria limestone (upper Raipur stage)
			Potnar, Sirsiguda, Bhanpuri, Barnji, Kadma, Deerguda, Devarapal, Tuaras, Dhuravaras, Jagdalpur, Majipal, and Kotamsar of Bastar district		Kanker (Jagdalpur stage) limestone (Indravati series)
		Madhya Pradesh	Sanama, Durjanpur, Nagod Maihar, Narairah, Banmore, Phalodi, Morak, Chibaura, Garhwar, Ramnagar, Degarhat, and Sejahata of Satna district		Upper Vindhyan limestone
			Palkua, Kachhra, and Maindri of Chhatarpur district		Tirohan limestone

(continued)

(continued)

S. no.	Limestone origin	Name of state	Locality	Cement plants and cement manufacturing	Remarks
			Bankuiyan, Bela, and Naubasta of Rewa district		
			Kailaras, Bakarpura, Jawahirgarh, and Garhi of Morena district		
			Kymore, Koilia, Murwar, Tikaria, Bistara, and Nimbahera of Katni-Jabalpur district		
		Odisha	Dungri, Sauntmal, Badmal, Behera, Kusumda, and Banjipalli of Sambalpur district		
		Karnataka	Hebbal Khurd, Waijal, Devapur, Salvadgi, Rajankollur, and Hartagi of Gulbarga district		Bhima series
			Kuchal, Talikot, Maileshwar, and Katgeri of Bijapur district		Bhima series
		Andhra Pradesh	Koilkuntla, Muttaluru, Allagadda, Tittapalle, Krishnapuram, Kottalle-Ramapalle, Nereducherla, Banganapalle, Auk, Satanikota, Bhogasamundram, Boyinacheruvapalle, Rammallukota, Brahmankotguru, Gargayapuram, Uyyalavid, Betamcherla, and Yannakondla of Kurnool district		Narji stage limestone of Kurnool system

(continued)

(continued)

S. no.	Limestone origin	Name of state	Locality	Cement plants and cement manufacturing	Remarks
			Ponnathota, Purvasungamanchepalle, Salelvariuppalapadu, Talamanchipatnam, Yerraguntla, Vellala, Koduru, and Nereducherla of Kadapa district		-do-
			Palnadu, Macherla, Nadikudi, Dachepalle Kesanupale, Gangulagunta, Gottipalle, and Kandlagunta of Guntur district		
			Jaggayapetta, Budavat, Jayantipuran, and Vedadri of Krishna district		
			Narella, Kamalapur, and Ganeshpalli of Karimnagar district		Sullavai series (Godavari Valley)
		Uttar Pradesh	Basuhar, Silpi, Markundi, Jamua, Belwa, Radauli, Makribari, Belach, Pataudh, Kadhaura, and Kajrahat of Mirzapur district		Rohtas stage limestone (Sone and Ghaghar Rivers)
		Bihar	Kajrahat, Chapri, Bajetoli, and Bhaunathpur of Palamu district		
			Jaradag, Chutia, Harriara Baulia, Jadunathpur, Ramdhira, Banjari Chunahatta, Gataihi, and Doma of Shahabad district		Rohtas stage limestone
		Rajasthan	Julmi, Mailo, Nimana, Deoli, Darra, and Ramgunjmandi of Kota district		

(continued)

(continued)

S. no.	Limestone origin	Name of state	Locality	Cement plants and cement manufacturing	Remarks
			Nimbahera, Keshavapur, Baldranha, Manpur, and Bhoikheda of Chhitorgarh district		
			Lakheri, Satur, Bundi of Bundi district		
			Phalodi, Maholi and Kela Devi of Sawaimadhopur district		
			Sojat of Jodhpur district and Bewar of Pali district		
4	Tertiary group	Haryana	Malla and Janpur (at Tundpathar, Kharag, Ramsar, and Sherla) of Ambala district	Bhupendra Cement Works	• Nummulitic limestone • Subathus of Eocene period • Miliolitic limestone of Pliocene and Pleistocene period
		Gujarat	Lakhpat and Ghuneri of Kutch district		Nummulitic limestone
		Gujarat	Porbandar and Dwarka	ACC Porbandar	Miliolitic limestone of Pleistocene period
		Gujarat	Adityana, Bharwada, and Bakherla (Ranavav) of Porbandar district Sikka and Dwarka, district Jamnagar	Digvijay cement and Saurashtra cement	Miliolitic limestone; cement and chemical grade
		Gujarat	Jar, Patanwao, Zirjuda, Uplata, and Paneli of district Gondal		Chemical grade limestone
		Gujarat	Veraval, Patan, Gorakhmudi, and Prachi of Gohilwad district		Miliolitic limestone

(continued)

(continued)

S. no.	Limestone origin	Name of state	Locality	Cement plants and cement manufacturing	Remarks
		Gujarat	Babarkot, Bhakodar, Vand, and Balana of Jaffrabad disrict	Narmada cement	-do-
		Gujarat	Kovaya, Adivi, Dholasa, and Khamandal of Amreli district	GCW cement works	-do-
		Gujarat	Tadkeshwar of Surat district	–	Nummulitic limestone
		Diu Island	Gulf of Cambay, Diu, and Malala, Nagoa, Dongarwadi, and Pavti		Miliolitic limestone
		Meghalaya	Therriaghat and Shella rivers of Khasi and Jaintia hills		
			Cherrapunjee (Mawmluh and Mawsmai) of Jaintia hills		
			Lumshnong, Garampani, Nongklih, and Syndai		
		Manipur	Ukhurul, Lambui, Sokpa, Shuganu, Toupokpi (Ukhrul), and Hungdung		
		Assam	Mikir Hills bordering Nawgong and Sibsagar districts, North Cachar, Lumbding, and Bokajan	CCI cement, Bokajan	
5	Jurassic system	Uttrakhand	Nilkanth, Pundras, Toli, Manikot Hill and Bhadsi of Garhwal district	–	Krol limestone belt (Upper Tal formations)
		Gujarat	Khavada Mahal, Abdasa of Kutch district	–	Middle Jurassic limestone
6	Cretaceous system	Tamil Nadu	Kallakudi and Kovandakurichi, Dalmiapuram of district Trichy	–	

(continued)

(continued)

S. no.	Limestone origin	Name of state	Locality	Cement plants and cement manufacturing	Remarks
			Ariyalur and Dalmiapuram	ACC cement (Sevalia Plant)	
		Madhya Pradesh	Semadhana and Mahuakhera of Sagar district		
		Andhra Pradesh	Kovvur (between Pangadi and Devrapalle) of West Godavari district	–	
7	Triassic system	J &K	Achibal-Bawan, Verinag-Zamalgam, Tserkar-Duroo, and Naupara of Anantnag district	–	Cement grade Triassic limestone
			Bandipura, Mansbal, Beru, Ajas, Gundi-Sunderkot, Sonmarg, Zojila, Khanan Bab Gund, Nadihal, Madar, and Brion of Bramulla district	–	
			Vihi valley between Wuyan and Khrew of Srinagar district	Wuyan cement plant	
8	Paleozoic group	Himachal Pradesh	Dundhiara and Mcleodganj-Dharamshala of Kangra district		
			Sataun, Nadi, Dadua, Bhatrog, and Kamraoo of Sirmour district	CCI, Rajban	
			Pungh, Dhanesri, and Jarel of Mandi district	–	Krol limestone
		Uttarakhand	Deoban, Kwanu, Chkrat, and Nagini of Dehradun district	–	

(continued)

(continued)

S. no.	Limestone origin	Name of state	Locality	Cement plants and cement manufacturing	Remarks
			Barkot, Bainsundhar, Ghoran hill, Ghorapitti hill, and Kutia ridge of Dehradun district	–	Krol limestone
			Jharipani, Sisoli, Shastradhara, Bhatta, and Hathipaon of Dehradun district	–	Upper Krol (Permian)
		Uttar Pradesh	Thal-Tijam of Pitthoragarh district	–	Calcareous zone
			Kosi, Bujan, Jaurasi, Baradhan Talla, and Batalghat of Nainital district	–	
		J &K	Bansoli of Kathua district	–	
9	Recent to subrecent period	Kerala	Vembanad lake area of Alleppey and Kottayam districts	–	Limeshell and Kankar limestone deposits
		Tamil Nadu	Rameshwaram	–	
		Andhra Pradesh	Pulicat lake area near Sulurpet of Nellore district	–	

Note Derived from NCCBM (1985)

Annexure B

Routine Condition Monitoring (RCM)

(Referenced in Sect. 2.6)

What is RCM?

Routine condition monitoring (RCM) is the application to test the equipment, on either a routine or continuous basis, for the performance and condition of machinery against acceptable standards. The technique not only enables machines to operate at optimum performance with improved reliability, but it is also used to provide trends in equipment deterioration that can be used to predict impending failures so that maintenance can be planned outside production periods, thereby reducing the number of breakdowns.

Methodology of RCM

The methodology of RCM involves the identification of a number of parameters that may indicate the health of a machine or any of its components. The acceptance values of these parameters for healthy functioning of the machine or components are established, and any undue variation of these parameters outside the acceptance range would indicate deterioration in the health of machines. Many parameters have been researched, studied, and identified for the purpose of RCM where (a) a *shock pulse* of vibration may be used for monitoring the health of bearings, (b) *debris content* in gear oil may be an indicator of health of gears, (c) the vibration and its mode monitored at different strategic points on a machine may give good idea regarding the overall health

A. K. Soni and P. Nema, *Limestone Mining in India*, Materials Horizons: From Nature to Nanomaterials, https://doi.org/10.1007/978-981-16-3560-1

of a machine, and (d) the exponential rise in temperature is a popular indicator of the health of electric motor. Thus, there are number of such parameters for mining machine.

RCM and Its Uses

The techniques of RCM may be used for multifarious purposes as described here:

Preindication of Failure: RCM may highlight deterioration of a machine's condition or performance and thereby help in the prevention of breakdown and the improvement of machine use. According to a field case study in a mine, machine breakdown and associated delays were reduced from 40.6 min per machine per shift to 15.8 min per machine per shift through the use of RCM. It should be appreciated that these improvements are indicated by the change in the value of the chosen parameter rather than absolute value, which is important for evaluating the health of a machine. Another example concerns the sudden increase in *debris content* in gear oil collected from a machine used in mine. The observation indicated that the debris content did not decrease even after changing the gear oil. Therefore, a defect was indicated that was rectified during the next weekly off day of work. While attending the defect, it was revealed that the machine was facing an impending breakdown, which could have been averted by a more timely action.

Commissioning Test: RCM is useful for the proper commissioning of electrical or mechanical contrivances. The technique may be used to ensure that the equipment has been properly assigned and is giving the desired performance. Such tests were conducted in the past for production of a face in an underground coal mine and gave improved performance of attended equipment (e.g., leakage in powered support and installation of spare stores in the vicinity of the working area).

Fault Diagnosis: RCM techniques may be used for diagnosing faults following unacceptable readings during a monitoring procedure.

Condition-Based Maintenance: RCM helps the maintenance staff to devote their time to those items of equipment whose conditions require attention (e.g., the greasing of bearings will contribute to economical use of both manpower and material).

Applications of RCM in Indian Geomining Conditions

The concept of RCM was first introduced in Indian mines by the Central Mining Research Institute (CSIR-CMRI, now CIMFR) in the coal mining industry with active support of the Director General of Mine Safety (DGMS). RCM-powered support is now accepted for working faces of coal mines in India where conditions are inseparably linked with the support performance and consequently the safety of people engaged at the work place. RCM use in limestone mines in the Indian mining industry is at a different stage compared to that of the government sector mines because most of the limestone-producing companies are owned by private sector organizations. The concept of RCM should be carried forward throughout the industry to receive full benefits of cost reduction.

Annexure C

Hydrogeochemical Analysis of Water

(Referenced in Sect. 4.2; Case Study 5)

Project-affected people (PAP) who makes use of BOP nala and amal nala water express their discontent from time to time that the water is being polluted due to mining operations and that the mine management should take necessary steps to curb water pollution. Palgaon, Taldhi, Awarpur, Bakardi, Naokari, Kusumbi (Lingandoh), Manoli Khurd, and Jamni are some en route villages of these nala from where periodic complaints of PAP's were received. Having noticed this and to check the pollution level, standard analytical methods [1] were used and analysis completed.

Using standard operating procedures (SOP), systematic field sampling was carried out during the post-monsoon season (November and December) when the aquifers are fully charged and water channels flow in a downstream direction. Representative samples were collected from different groundwater locations in dug wells, bore wells, and nala around the Naokari and the Manikgarh limestone mining areas. In all, 31 samples were collected from both the shallow and deep aquifers of the Penganga limestone, shale, Kamthi sandstone and Barakar sandstone (see Fig. C1). Surface water samples are collected considering the upstream and downstream direction of the water channel flowing through these mine areas under study. Some points based on hydrogeochemical analysis of water are as summarized here:

1. Consider that groundwater contained in an aquifer of any given area has a unique chemistry that is acquired as a result of chemical alteration of meteoritic water recharging the system [4, 7]. The pH of the groundwater in the study area (bore well and dug well) lies in the range of 6.9–8.6, EC is in range of in 184–4280 μS/m, the total hardness (TH) varies from 184–1589 mg/L, total dissolved solids (TDS) values range from 192–2739 mg/L, the HCO_3 varies from 160–

A. K. Soni and P. Nema, *Limestone Mining in India*, Materials Horizons: From Nature to Nanomaterials, https://doi.org/10.1007/978-981-16-3560-1

Fig. C1 Drainage network and sampling locations of study area

560 mg /L, the sulphate (So_4) values range from 15–204 mg/L, sodium (Na^+) values range from 16–255 mg/L, and fluoride (F^-) influence is of the order of 0.5–6.7 mg/L (Tables C1 and C2). Similarly, the surface water pH is since greater than 7. Therefore, its nature can be categorized as basic and hard as the TDS values are higher (SW1–SW3 points, as shown in Table C2).

Table C1 Chemical composition of groundwater from Naokari and Manikgarh limestone mine areas (dug well)

Village Name	Well ID	Aquifer	pH	EC μS/m	Temp (°C)	TDS	TH	Ca^{2+}	Mg^{2+}	TA	CO_3^{2-}	HCO_3^-	Cl^-	Na^+	K^+	Fe^{2+}	F^-	SO_4^{2-}	NO_3^-
Kadoli	DW1	Penganga Shale, Lst	8.5	664	27	425	320	53	45	224	64	160	64	29	1	0.28	1.12	22	44
Dhamamgaon	DW2	Penganga Shale, Sst	8.4	2470	29	1581	1136	197	155	328	32	296	316	51	3	0.3	0.86	66	421
Bibigaon	DW3	Penganga Shale, Lst	8.5	1700	26	1088	784	133	108	252	32	220	258	56	4	0.46	0.75	124	204
Nanda phata	DW4	Penganga Shale, Lst	7.2	1587	25	1016	760	104	120	380	0	380	190	59	4	0.38	1.15	85	159
Palgaon	DW6	Penganga Lst	7.5	2250	27	1440	872	291	35	200	0	200	260	72	3	0.48	1.25	204	253
Awarpur	DW7	Penganga Shale, Lst	8.4	1071	26	685	328	58	44	380	32	348	80	100	3	0.4	1.2	64	35
Gadegaon	DW8	Sst	8.5	2800	26	1792	656	43	132	608	48	560	316	255	53	0.28	4.23	158	204
Talodi	DW9	Penganga Shale, Lst	7.6	1915	26	1226	336	45	54	512	0	512	190	226	38	0.28	6.57	111	155
Khairgaon	DW17	Penganga Shale	7.6	1334	27	854	420	101	40	348	0	348	110	133	4	0.41	1.34	86	146

(continued)

Table C1 (continued)

Village Name	Well ID	Aquifer	pH	EC μS/m	Temp (°C)	TDS	TH	Ca^{2+}	Mg^{2+}	TA	CO_3^{2-}	HCO_3^-	Cl^-	Na^+	K^+	Fe^{2+}	F^-	SO_4^{2-}	NO_3^-
Injapur	DW5	Penganga Shale	7.6	1558	27	997	496	90	65	380	0	380	104	67	129	0.52	1	84	248
Bombezari	DW10	Deccan Basalt	7.5	685	25	438	304	45	46	308	0	308	28	18	2	0.52	1.1	15	31
Naokari Khurd	DW11	Penganga Lst	7.5	1727	26	1150	656	163	60	280	0	280	178	28	26	0.28	0.9	172	248
Guwariguda	DW12	Penganga Lst	7.5	721	25	461	336	64	42	280	0	280	36	16	4	0.28	0.5	22	58
Belampur	DW13	Penganga Lst	7.6	883	25	565	420	94	44	280	0	280	70	22	8	0.3	1.2	21	97
Gadchandur	DW14	Penganga Shale, Lst	7.4	1159	27	742	540	112	62	320	0	320	146	41	3	0.48	0.9	40	58
Manoli khurd	DW15	Penganga Shale, Lst	7.6	1015	26	650	436	83	55	280	0	280	66	38	3	0.46	1.3	45	137
Nagrala	DW16	Deccan Basalt	6.9	1131	27	724	568	120	64	224	0	224	110	16	4	0.3	1.1	63	168

Note All values are in mg/L unless stated, except pH, EC in μS/m Temp in °C
TH total hardness, *TA* total alkalinity, *DW* dug well, *BW* bore well, *SW* Surface water, *Lst* Limestone, and *Sst* Sandstone

Table C2 Chemical composition of groundwater from Naokari and Manikgarh limestone mine areas (bore well and surface water)

Village name	Well ID	Aquifer	pH	EC μS/m	Temp (°C)	TDS	TH	Ca^{2+}	Mg^{2+}	TA	CO_3^{2-}	HCO_3^-	Cl^-	Na^+	K^+	Fe^{2+}	F^-	SO_4^{2-}	NO_3^-
Wadegaon	BW1	Penganga Shale	7.3	1481	25	948	608	69	105	428	0	428	136	90	6	0.5	1.47	41	173
Asan khurd	BW2	Penganga Shale	7.1	1853	24	1186	560	86	83	432	0	432	184	175	1	0.62	6.64	78	253
Bori Nawegaon	BW3	Penganga Shale, Lst	7.3	827	25	529	360	70	44	296	0	296	48	44	4	0.52	6.35	25	80
Sonurli	BW4	Kamthi Sst	7.2	730	26	467	364	109	22	280	0	280	50	15	4	0.42	0.69	20	66
Palgaon	BW5	Penganga Shale, Lst	7.2	4280	26	2739	1536	336	167	376	0	376	834	140	126	0.5	1.4	168	284
Pipalgaon	BW6	Penganga Shale, Lst	7.4	1430	25	915	480	109	50	332	0	332	100	35	10	0.38	1.06	61	128
Hirrapur	BW7	Penganga Shale	7.3	1460	25	934	420	51	70	508	0	508	76	113	3	0.36	1.43	46	66
Antargaon	BW8	Penganga Shale	8.4	1086	26	695	392	45	67	460	56	404	70	157	1	0.28	1.38	25	84
Kavthala	BW9	Kamthi Sst	7.2	1440	25	922	648	176	50	320	0	320	130	63	6	0.26	1.28	44	217
Nokhari BZ-ACW	BW11	Penganga Shale, Lst	7.4	896	25	573	452	120	36	292	0	292	48	21	3	0.24	1.21	78	62

(continued)

Table C2 (continued)

Village name	Well ID	Aquifer	pH	EC μS/m	Temp (°C)	TDS	TH	Ca^{2+}	Mg^{2+}	TA	CO_3^{2-}	HCO_3^-	Cl^-	Na^+	K^+	Fe^{2+}	F^-	SO_4^{2-}	NO_3^-
Bakardi	BW14	Penganga Shale	8.6	1092	26	699	536	86	77	268	40	228	96	33	1	0.34	1.1	41	168
Kusumbi (Lingandoh)	BW10	Deccan Basalt	7.4	623	25	417	597	235	66	268	0	268	104	71	2	0.3	0.91	58	248
Manoli khurd	BW12	Penganga Shale	8.4	544	26	348	184	11	37	208	48	160	30	72	7	0.24	0.68	18	35
Jamni	BW13	Penganga Shale	7.2	2370	25	1571	832	179	92	304	0	304	232	48	109	0.28	1.14	83	487
Amal nala Dam water*	SW1		8.2	319	26	192	160	36.1	17	160	2.3	157.6	20	14.2	0.4	0.32	0.83	7.5	5.51
Downstream of Bop Nala*	SW2		8.3	657	26	408	256	57.5	27.2	228	4.6	223.3	30	22.4	0.8	0.28	0.96	90.8	8.5
Upstream of Bop Nala*	SW3		8.4	680	26	416	252	57.7	26.2	212	3.8	173.6	32	24.3	0.9	0.41	1.1	94.2	6.13

Note All values are in mg/L unless stated, except pH, EC in μS/m, Temp in °C
TH total hardness, *TA* total alkalinity, *BW* bore well, *SW* surface water, *Lst* =limestone, and *Sst* = sandstone

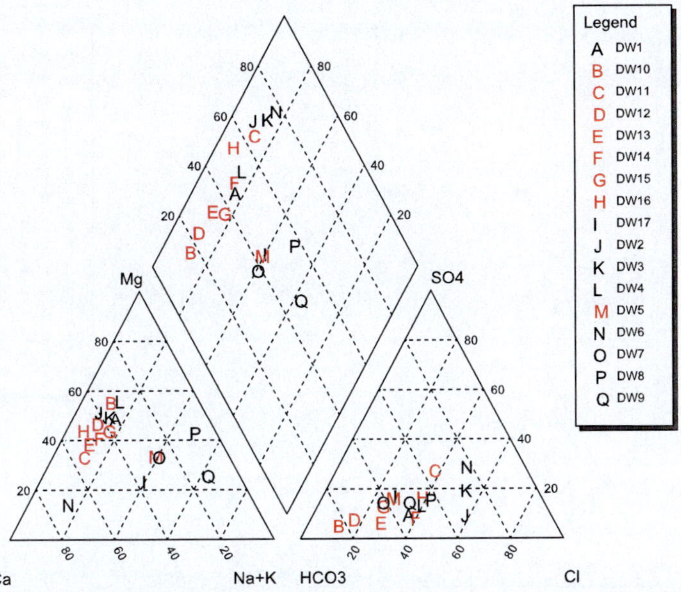

Fig. C2 Piper diagram of groundwater samples from Naokari and Manikgarh limestone mine areas (post-monsoon 2010 dug well samples)

2. Using a *Piper trilinear diagram* [12] and Aquachem 4.0 software, the determination of hydrochemical facies was carried out. According to the Piper diagram, the majority of the groundwater samples belong to Type I: Ca–Mg–HCO_3 where alkaline earth exceeds the alkali, and Type III and Type IV: mixed CaMgCl and mixed Ca–Na–HCO_3 where alkalis exceed alkaline earths and weak acids exceed strong acids. The water types show that Mg^{2+} is the dominant cation followed by Ca^{2+} and Na^+ (Figs. C2 and C3).
3. Using *Gibbs variation diagrams* [8] analysis of the mechanism of rock–water interaction in the evolution of the groundwater quality has been understood. The chemical data of groundwater samples of the study areas was plotted in Fig. C4a–d. The diagrams represent the interrelationships of TDS [$Na^+ + K^+ / (Na^+ + K^+ + Cl^-)$] as well as TDS [$Cl^-$ ($Cl^- + HCO_3{}^-$)]. The distribution of sample points in Gibbs diagram exhibits that chemical weathering of rock-forming minerals is the main factor causing the evolution of chemical composition of groundwater in the study area, which is later influenced by anthropogenic activities. The ratio of cations Na^+: ($Na^+ + Ca^+$) and anions Cl^-: ($Cl^- + HCO_3{}^-$) are spread from the rock domain towards the zone of precipitation dominance, indicating the rock–water interactions as a major source of dissolved ions in the groundwater. In addition, anthropogenic sources, such as the application of fertilizers and excess watering for agriculture, are also active to govern the groundwater chemistry in the study area. The anthropogenic activities lead to increased Na^+ and Cl^- ions, which in turn increases TDS [13]. From the Gibbs diagrams, it is clear that

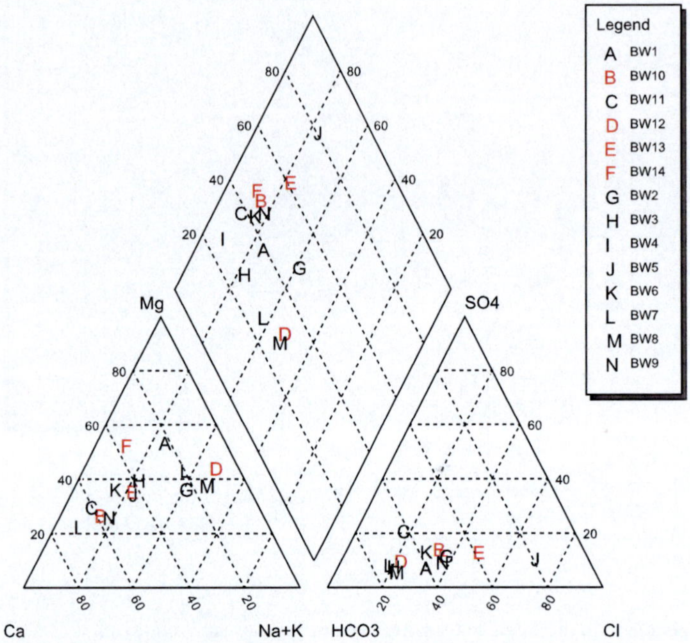

Fig. C3 Piper diagram of groundwater samples from Naokari and Manikgarh limestone mine areas (post-monsoon 2010 bore well samples)

maximum samples lie in the *rock dominance zone* and only few samples are in the *evaporation dominance zone*.

4. A *Wilcox plot*, which is also known as the *U.S. salinity diagram* [15], has been used to quickly determine the viability of water for irrigation purposes. The Wilcox plot is a simple scatter plot of sodium hazard using the sodium-adsorption ratio (SAR) on the Y-axis versus the salinity hazard (conductivity) on the X-axis. The salinity (conductivity) is plotted by default in a log scale. This plot has conductivity (us/cm) where C1: low (0–249), C2: medium (250–749), C3: high (750–2249), and C4: very high (2250–5000). The sodium (SAR) values are divided into the categories where S1: low, S2: medium, S3: high, and S4: very high. According to this, the groundwater samples from the Naokari and Manikgarh limestone mine areas have medium-to-very high salinity hazard (Figs. C5 and C6), mostly falling in the high-salinity zone.

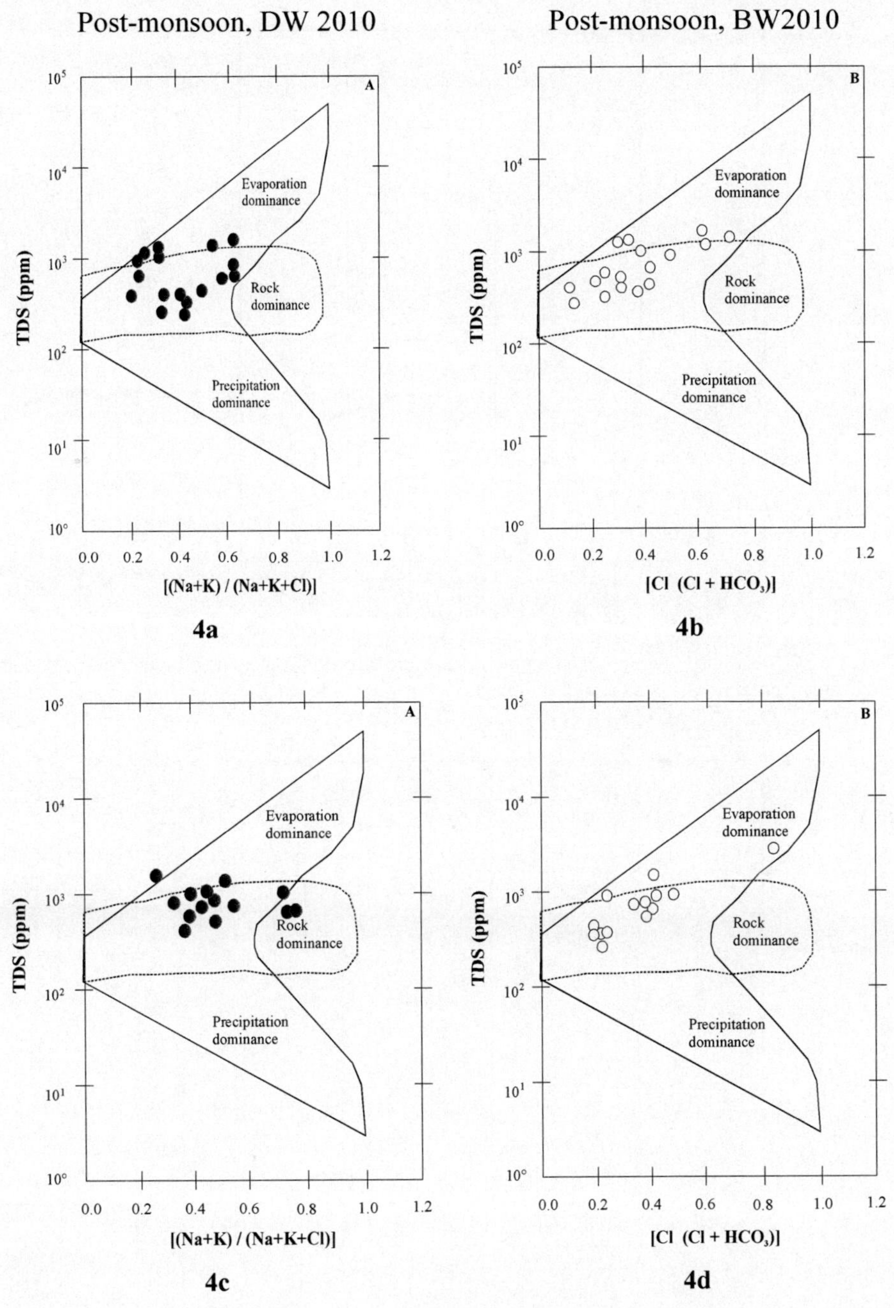

Fig. C4 a–d Gibbs diagram for groundwater samples from Naokari and Manikgarh limestone mine areas

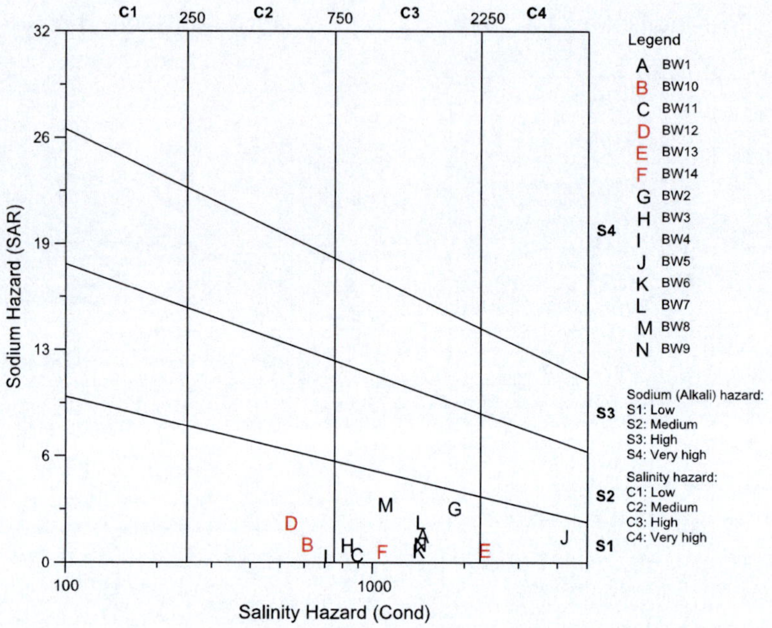

Fig. C5 U.S. salinity diagram of groundwater samples from Naokari and Manikgarh limestone mine areas (post-monsoon; dug well samples)

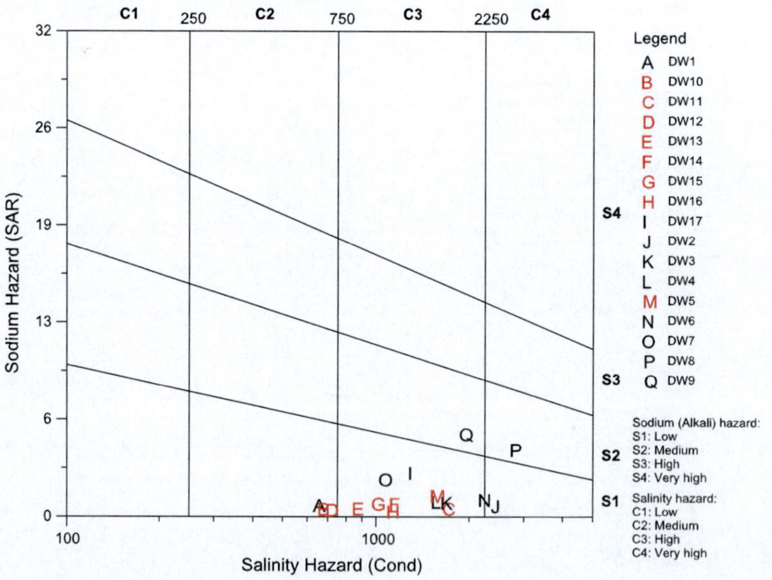

Fig. C6 U.S. salinity diagram of groundwater samples from Naokari and Manikgarh limestone mine areas (post-monsoon; bore well samples)

Table C3 Other water quality results from Naokari and Manikgarh limestone mine areas*

• **SAR Results**: Most of the water samples around the mining areas fall within high salinity hazards C2–C3, S1 field. Three locations of dug wells and two locations from bore wells are in a very high-salinity zone
• **% Na Results**: The sodium percentage in the study area ranges from 6.62 to 61.76% in dug well and 9.43–47.96% in bore well waters
• **RSC Results**: The RSC values of nearly all of the water samples from dug wells and bore wells (except one dug well in the village of Talodhi) in the study area are suitable for irrigation
• **CR Results**: All of the water samples of the study area having CR values ≤ 1 indicate that water in the area is good for irrigation ([except two locations in DW8 (Gadegaon) and BW-5 (Palgaon)]
• **KR Results**: Evaluated KR results of the groundwater show that water of the study area is suitable for irrigation (Talodhi village and Awarpur villages from Naokari limestone mine area)
• **PI Results**: In the study area, PI values of the post-monsoon season fall in class I (dug well: 17.827–77.051; bore well: 33.173–85.138). Hence, on the basis of computed PI values that represent the class I of Doneen (1964), it can be concluded that the groundwater from Naokari and Manikgarh limestone mine areas is suitable for irrigation
• **SSP Results**: All of the SSP values of pre- and post-monsoon dug wells as well as borewell water samples collected from Naokari and Manikgarh limestone mine areas lie below the threshold limit of SSP $\leq$ 50 (Sonurli village point BW4 and the village of Talodhi of the Naokari limestone mine area)

Source Khond [11]
NoteSAR sodium absorption ratio, *RSC* residual sodium carbon, *RSBC* residual sodium bicarbonate, *CR* corrosivity ratio, *SSP* soluble sodium percentage, *KR* Kelley's ratio, and *PI* permeability index
*For different values of parameter please refer Table 5a, b

5. The USLS classification from both mines shows that water of the area belongs to C2-S1 and C3-S1 classes, which indicate water of medium to high salinity–low sodium type, which can be used for irrigation with very little danger of exchangeable sodium.
6. The results of the obtained sodium adsorption ratio (SAR), percentage sodium (%Na), residual sodium carbonate (RSC), corrosivity ratio (CR), residual sodium bicarbonate (RSBC), Kelley's ratio (KR), permeability index (PI), and soluble sodium percentage (SSP) needed to know the suitability of water for various purpose has also been evaluated (see Table C3), and it was found that the water from limestone mining areas can be utilized for irrigation purposes.

7. The Naokari and Manikgarh are limestone mines, and the contents of Ca^{++}, Mg^{++}, and HCO_3^- are principally controlled by a carbonate mineral–water equilibrium. Appelo and Postma [2] have propounded that HCO_3^- originates from the dissolution of carbonate rocks and emission of CO^2 gas. In such water, the total hardness and total alkalinity remain an issue from a pollution perspective.

Based on this analysis, it is determined that villages in and around these limestone mining areas are inflicted with fluoride hazards. Most of the water can be used for irrigation purposes but are not fit for drinking purposes. If the water of the area is to be used for domestic purpose (miscellaneous uses), it should be treated first.

Defluoridation and *artificial recharge* are common practices for the prevention and control of fluorosis. The *Nalgonda technique* is a simple method of removing excess fluoride that is based on the principle of adsorption where ion exchanges can be used [3, 16]. Artificial recharge (a low-cost natural and best management solution) reduces the fluoride content considerably [6, 9, 10]. Check dams constructed for promoting artificial recharge also result in the decrease of the fluoride content. In this case study, the limestone mining area F^- content can be brought down from 3.10 mg/l to below the recommended limit. These management solutions—regarding with the nutritional supplementation of the water in the area—are insufficient for normal drinking. However, drinking water can be supported by adding calcium, vitamin C, and antioxidants. This will be beneficial for the prevention of fluorosis [14].

In general, water has the characteristic of being neutral with high electrical conduction (EC) values. The quality of water is termed as *affected* when higher TDS and fluoride content is observed. Therefore, water management practices need to evolve, which is a challenging task.

To evaluate the level of pollution around a limestone mining area, the analytical values have been compared to the World Health Organization [17] and the Bureau of Indian Standards [5] values of drinking water standards (Table C4). These guidelines of BIS and WHO indicated that the surface water channels of the area have higher fluoride content and hardness content, exceeding the maximum permissible level set by the BIS and WHO standards [11].

Table C4 Comparison of the chemical parameter of groundwater in study area with WHO [17] and BIS [5] drinking water standards

Parameter (units)	WHO standard [17]		BIS [5]		Post-monsoon dug well		Post-monsoon bore well	
	Maximum acceptable limit	Maximum allowable limit	Maximum acceptable limit	Maximum allowable limit	Min	Max	Min	Max
pH	6.5	8.5	6.5	8.5	6.9	8.5	7.1	8.6
EC (μS/m)	–	–	–	–	664	2800	544	4280
TDS	500	1500	500	2000	425	1792	348	2739
TH	100	500	300	600	304	1136	184	1536
Ca^{++}	75	200	75	200	43	291	11	336
Mg^{++}	50	150	30	100	35	155	22	167
Na^{+}	50	200	50	200	16	255	15	175
K^{+}	–	200	–	–	1	129	1	126
HCO_3^{-}	30	150	30	150	160	200	160	508
Cl^{-}	200	600	200	1000	28	316	30	834
SO_4^{--}	200	400	200	400	15	204	18	168
NO_3^{-}	–	45	–	45	31	421	35	487
F^{-}	1.0	1.5	1.0	1.5	0.53	6.57	0.7	6.6

Note All values are in mg/L unless stated, except pH

Geochemical classifications, ion-exchange processes, mechanisms controlling groundwater chemistry, sodium hazards, and Ca $^{2+}$: Mg^{2+} ratios show that most of the water samples in and around the mining areas can be used for irrigation without any hazards (Table C5a, b). The water quality is unsafe for drinking purposes, but it

Table C5 Irrigation quality results of groundwater samples from Naokari and Manikgarh limestone mine areas (dug well)

(a)

Sample No	SAR	% Na	RSC	RSBC	CR	SSP	KR	PI
1	0.250	16.86	−1.592	−0.0223	0.079	16.860	0.198	37.866
2	0.233	9.22	−16.663	−4.978	0.680	9.226	0.098	17.827
3	0.309	14.05	−10.849	−3.030	0.467	14.055	0.156	24.140
4	0.330	15.02	−8.832	1.0386	0.584	15.052	0.170	28.717
5	0.464	38.7	−3.609	1.7372	0.370	38.708	0.296	42.424
6	0.375	15.56	−14.122	−11.242	0.483	15.569	0.18	24.072
7	0.852	40.46	0.255	2.809	0.260	40.462	0.667	62.025
8	1.538	48.9	−2.227	7.032	1.469	48.907	0.853	58.606
9	1.900	61.76	1.704	6.146	0.835	61.764	1.470	77.051
10	0.159	12.15	−0.981	2.802	0.072	12.153	0.129	44.474
11	0.168	12.59	−8.480	−3.544	0.495	12.592	0.093	23.519
12	0.134	10.71	−2.059	1.3956	0.087	10.719	0.104	38.644
13	0.165	12.26	−3.720	−0.101	0.148	12.263	0.115	33.443
14	0.272	14.82	−5.444	−0.344	0.349	14.823	0.166	32.661
15	0.280	16.63	−4.076	0.4475	0.165	16.638	0.190	36.779
16	0.103	6.62	−7.581	−2.316	0.210	6.624	0.061	21.860
17	1.00	41.41	−2.626	0.663	0.355	41.410	0.694	57.904

(b)

Sample No	SAR	% Na	RSC	RSBC	CR	SSP	KR	PI
1	0.250	16.86	−1.592	−0.0223	0.079	16.860	0.198	37.866
2	0.233	9.22	−16.663	−4.978	0.680	9.226	0.098	17.827
3	0.309	14.05	−10.849	−3.030	0.467	14.055	0.156	24.140
4	0.330	15.02	−8.832	1.0386	0.584	15.052	0.170	28.717
5	0.464	38.7	−3.609	1.7372	0.370	38.708	0.296	42.424
6	0.375	15.56	−14.122	−11.242	0.483	15.569	0.18	24.072
7	0.852	40.46	0.255	2.809	0.260	40.462	0.667	62.025
8	1.538	48.9	−2.227	7.032	1.469	48.907	0.853	58.606
9	1.900	61.76	1.704	6.146	0.835	61.764	1.470	77.051
10	0.159	12.15	−0.981	2.802	0.072	12.153	0.129	44.474

(continued)

Table C5 (continued)

(a)

Sample No	SAR	% Na	RSC	RSBC	CR	SSP	KR	PI
11	0.168	12.59	−8.480	−3.544	0.495	12.592	0.093	23.519
12	0.134	10.71	−2.059	1.3956	0.087	10.719	0.104	38.644
13	0.165	12.26	−3.720	−0.101	0.148	12.263	0.115	33.443
14	0.272	14.82	−5.444	−0.344	0.349	14.823	0.166	32.661
15	0.280	16.63	−4.076	0.4475	0.165	16.638	0.190	36.779
16	0.103	6.62	−7.581	−2.316	0.210	6.624	0.061	21.860
17	1.00	41.41	−2.626	0.663	0.355	41.410	0.694	57.904

NoteSAR sodium absorption ratio, *RSC* residual sodium carbon, *RSBC* residual sodium bicarbonate, *CR* corrosivity ratio, *SSP* soluble sodium percentage, *KR* Kelley's ratio, and *PI* permeability index

can be made useful for domestic purposes after treatment and disinfection.

References

1. Rand MC, Greenberg AE, Taras MJ (1998) Standard methods for the examination of water and waste water. American Public Health Association (APHA), Washington, DC, pp 6–187
2. Appelo CAJ, Postma D (1996) Geochemistry, groundwater and pollution. A.A. Balkema Publication, Rotterdam, p 536
3. Ayoob S, Gupta AK (2006) Fluoride in drinking water: a review on the status and stress effects. Crit Rev Environ Sci Technol 36(6):433–487
4. Back W (1961) Technique is for mapping of hydro chemical facies. USGS Prof Paper 424D, Reston, pp 380–382
5. BIS (2003) Indian standard: specification for drinking water (IS: 10500), Bureau of Indian Standard (BIS), New Delhi, pp 1–5
6. Dharashivkar AP, Khond MV, Murkute YA (2014) Preliminary report of fluoride contamination in groundwater from PG-2 watershed of Penganga River Basin, Korpana Taluka, Chandrapur District, Maharashtra. Gondwana geological magazine, Special vol 14, pp 161–166
7. Drever JI (1982) the geochemistry of natural waters: surface and groundwater environments. Prentice Hall, India Pvt. Ltd., p 182
8. Gibbs RJ (1970) Mechanism controlling world's water chemistry. Science 170:1080–1090
9. GSDA (2012) Impact assessment on cluster based sustainability by injection recharge in parts of WRD watershed of Tehsil Warora, District Chandrapur, Groundwater Survey Development Agency (GSDA), Government of Maharashtra, Technical report (Unpublished)
10. Jacks G, Bhattacharya P, Chuodhary V, Singh KP (2005) Control on the genesis on some high-fluoride groundwater in India. Appl Geochem 20:221–228

11. Khond MV (2014) Hydrogeochemical characterization of groundwater around Naokari and Manikgarh Limestone Mines, Chandrapur District, Maharashtra, Rashtrasant Tukadoji Maharaj Nagpur University (RTMNU), Nagpur, Ph.D. thesis (Unpublished), p 210
12. Piper AM (1944) A graphical procedure in the geochemical interpretation of water analyses. Am Geophys Union Trans 25:914–923
13. Subba Rao N, Nirmala IS, Suryanarayan K (2005) Groundwater quality in a coastal area: a case study from Andhra Pradesh, India. Environ Geol 48:543–550
14. UNICEF (1999) State-of-art report on the extent of fluoride in the drinking water and the resulting endemicity in India, United Nation's India Children Education Fund (UNICEF), New Delhi, pp 105–107.
15. USLS (1954) Diagnosis and improvement of saline and alkali soils, U.S. department of agriculture handbook. In: Richards LA (ed) U.S. Salinity Laboratory Staff, (USLS), p 60
16. Waghmare SS, Tanvir A (2015) Fluoride removal from water by various techniques: a review. Int J Innov Sci Eng Technol (IJISET) 2(9):2348–7968. (www.ijiset.com) ISSN
17. WHO (2004) Guideline for drinking water quality recommendation. World Health Organization (WHO) 1:1–4

Annexure D

Statuary Compliance Under Indian Acts and Rules for Limestone Sector (Referenced in Sect. 3.2.5)

S. No.	Concerned act and rules	Procedural requirements	Type of statuary compliance
Mines and Mining			
1.	Mines Act-1955	Opening of mine	Intimation in Form-I
2.	Mines and Minerals (development and regulation) Act (MMDR), 1957 and its amendments	Grant of lease	Permission
3.	MMDR Act, 1957	Execution of lease	Permission
4.	MMDR Act, 1957	Renewal of lease	Permission
5.	MMDR Act, 1957	Inclusion of new mineral grant and its execution	Permission
6.	MMDR Amendment Act-2015	Grant of extension of lease	Permission
7.	MMDR Amendment Act, 2015	Execution of extension of lease deed	Permission
8.	MMDR Act, 2015	Payment of DMF @ 30% of royalty	Statutory payment (under MTS)

(continued)

A. K. Soni and P. Nema, *Limestone Mining in India*, Materials Horizons: From Nature to Nanomaterials, https://doi.org/10.1007/978-981-16-3560-1

(continued)

S. No.	Concerned act and rules	Procedural requirements	Type of statuary compliance
9.	MMDR Act, 2015	Payment of NMET @ 2% of royalty	Statutory payment (under MTS)
10.	Minerals Concession Rules (MCR), 1960 and MCR, 2016	Approval of mining plan or mining scheme	Permission and approval
11.	MCR, 2016	Payment of royalty	Statutory payment (under MTS)
12.	Minerals Conservation and Development Rules (MCDR), 2017	Monthly return in Form F-1	Filing and returns
13.	MCDR, 2017	Annual return in Form G-1	Filing and returns
14.	MCDR, 2017	Monthly return in Form L	Filing and returns
15.	MCDR, 2017	Annual return in Form M	Filing and returns
16.	MCDR, 2017	Details of exploration	Intimation and reporting
17.	MCDR, 2017	Application for registration as a dealer	Registration and license
18.	Welfare & Cess Act, 1973	Production and dispatch of limestone	Filing and returns in Form-D
19.	Welfare & Cess Act, 1973	Monthly receipt of limestone	Filing and returns in Form-E
20.	Payment of Wages Act (Mines), 1956	Number of persons employed or number of working days (average), etc.	Filing and returns in Form-V
21.	Minimum Wages (Central) Rules, 1950	Number of persons employed or average number of working days, etc.	Filing and returns in Form-III
22.	MCR and MMR	Intimation of appointment to IBM and DGMS	Intimation
23.	Metalliferrous Mines Regulation (MMR), 1961	Quarterly return of mines data, mandays/workdays (as per MMR 1961), payment, production etc	Filing and returns in Form-II
24.	Metalliferrous Mines Regulation (MMR), 1961	Annual return of mines data, mandays/workdays (as per MMR, 1961), payment, production, etc.	Filing and returns in Form-III
25.	Mines Rules, 1955	Return of reportable accidents	Filing and returns in Form-J and Form-K
26.	Mines Rules, 1955	Medical examination	Filing and returns in Form-T

(continued)

(continued)

S. No.	Concerned act and rules	Procedural requirements	Type of statuary compliance
27.	Indian Explosives Act (IEA), 1884 and Indian Explosives Rule (IER), 2008	Monthly return for manufacturing and use of ANFO	Filing and returns (RE-2)
28.	IEA, 1884 and IER, 2008	Quarterly return for use of explosives	Filing and returns (RE-7)
29.	Explosive Rules, 2008	Explosive license and magazine for storage of explosives	License
30.	Explosive Rules, 2008	Explosive license and new explosive van\for explosive transportation	License
31.	Explosive Rules, 2008	Inland bulk use of highly inflammable petrol or petroleum products in mines and license for high speed diesel (HSD) storage, use, and transport in mines	License
32.	Metalliferrous Mines Regulation (MMR), 1961	Deep hole drilling and blasting	Permission
33.	Metalliferrous Mines Vocational Training Rules (MVTR), 1966	Vocational training (VT) centre	Permission
34.	Central Electricity Regulation, 2010	Use of HT installations	Permission
35.	Indian Electricity Rules, 1956	Annual return of electrical installations	Filing and returns in Form-X
36.	Central Electricity Regulation, 2010	Electrical apparatus and installations	Permission and approval
Environment			
37.	EIA Notification, 1994/2006	Environmental clearance	Permission
38.	EIA Notification, 1994/2006	Environmental clearance	Compliance
39.	Water (Prevention and Control of Pollution) Act, 1974	Consent to operate	Permission
40.	Water (Prevention and Control of Pollution) Act, 1981	Consent to operate	Permission and compliance

(continued)

(continued)

S. No.	Concerned act and rules	Procedural requirements	Type of statuary compliance
41.	Hazardous Waste (Management, Handling, and Transboundary Movement) Rule, 2008	Consent to operate with compliance	Permission and compliance
42.	Central Ground Water Board (CGWB)/Central Groundwater Authority (CGWA)	Groundwater withdrawal	Permission and compliance (NOC)
43.	Forest Conservation Act, 1980	Clearances for operating in forest area (i.e., wildlife preservation)	Permission, compliance, and approval